Glencoe Science

BIOLOGY

The Dynamics of Life

Reinforcement and Study Guide

 Glencoe

New York, New York   Columbus, Ohio   Chicago, Illinois   Peoria, Illinois   Woodland Hills, California

A GLENCOE PROGRAM

BIOLOGY: THE DYNAMICS OF LIFE

Glencoe Online
SCIENCE

Visit the Glencoe Science Web site
bdol.glencoe.com

You'll find:
Standardized Test Practice, Interactive Tutor, Section and Chapter Self-Check Quizzes, Online Student Edition, Web Links, Microscopy Links, WebQuest Projects, Internet BioLabs, In the News, Textbook Updates, Teacher Bulletin Board, Teaching Today

and much more!

 Glencoe

The *McGraw-Hill* Companies

Send all inquiries to:
Glencoe/McGraw-Hill
8787 Orion Place
Columbus, OH 43240-4027

ISBN 0-07-860232-7

Printed in the United States of America.

5 6 7 8 9 10   009   08 07 06 05

Contents

To the Student

This *Reinforcement and Study Guide* for **Biology: The Dynamics of Life** will help you learn more easily from your textbook. Each textbook chapter has four study guide pages of questions and activities for you to complete as you read the text. The study guide pages are divided into sections that match those in your text. Each BioDigest in your textbook has two study guide pages to complete.

You will find that the directions in the *Reinforcement and Study Guide* are simply stated and easy to follow. Sometimes you will be asked to answer questions. Other times, you will be asked to label a diagram or complete a table. By completing the study guide, you will gain a better understanding of the concepts presented in the text. These sheets also will prove helpful when studying for a test.

Before you begin your work, read the Study Skills section at the back of the book. The Study Skills section will help you

- improve your reading skills.
- improve your vocabulary skills.
- learn from visuals.
- make and understand idea maps.

Biology: The Study of Life

Reinforcement and Study Guide

Section 1.1 What is biology?

In your textbook, read about the science of biology.

Answer the following questions.

1. What is the primary focus of all biological studies?

_____

_____

2. What is meant by the statement, "Living things do not exist in isolation"?

_____

_____

_____

In your textbook, read about why biologists study the diversity of life.

For each of the statements below, write <u>true</u> or <u>false</u>.

_____ **3.** People study biology only if they are planning to become biologists.

_____ **4.** By studying biology, you can better appreciate the great diversity of species on Earth and the way each species fits into the dynamic pattern of life on the planet.

_____ **5.** The study of biology includes the investigation of interactions among species.

In your textbook, read about the characteristics of living things.

Complete each statement.

6. To be considered _____ , something must exhibit all of the _____ of life.

7. _____ is another word for "living thing."

8. Every living thing, from simple, single-celled organisms to complex, multicellular plants and animals, is made up of parts that function together in orderly living _____ .

Read each of the following statements. If it describes the process of reproduction, write <u>yes</u>. If not, write <u>no</u>.

_____ **9.** New leaves appear on a tree in spring.

_____ **10.** An amoeba divides in half.

_____ **11.** A bean plant produces seeds in long pods.

_____ **12.** Pollen grains are released from a flower.

_____ **13.** A sea star produces a new arm after losing one to a predator.

Circle the letter of the choice that best completes the statement.

14. A species is defined as a group of similar-looking organisms that
  a. undergo similar developmental changes.   **b.** can interbreed.
  c. can interbreed and produce fertile offspring.   **d.** reproduce in the same way.

15. Every organism begins life as a(n)
  a. embryo.   **b.** single cell.   **c.** nucleus.   **d.** fertilized egg.

16. A corn plant producing ears of corn is an example of
  a. growth.   **b.** reproduction.   **c.** development.   **d.** all of these.

17. If members of a species fail to reproduce successfully, the species
  a. will eventually become extinct.   **b.** will not develop normally.
  c. will evolve into a new species.   **d.** will remain unchanged.

Complete the table below by checking the correct column for each example.

Example	Stimulus	Response
18. The recess bell ringing at an elementary school		
19. Your mouth watering at the sight of food on a plate		
20. A sudden drop in air temperature		
21. A flu virus entering your body		
22. Getting butterflies in your stomach before giving a speech		

Answer the following questions.

23. Explain the concept of homeostasis.

_____

_____

24. What is an adaptation?

_____

_____

25. What is evolution?

_____

Chapter 1 **Biology: The Study of Life,** continued

In your textbook, read about observing and hypothesizing.

Answer the following questions.

1. What is meant by *scientific methods*? _____

_____

2. What is a hypothesis? _____

_____

3. How is a hypothesis tested? _____

In your textbook, read about experimenting.

For each item in Column A, write the letter of the matching item in Column B.

Column A	Column B
_____ **4.** A procedure that tests a hypothesis by collecting information under controlled conditions	**a.** dependent variable
_____ **5.** In an experiment, the group that is the standard against which results are compared	**b.** experimental group
_____ **6.** In an experiment, the group in which all conditions are kept the same except for the one being tested	**c.** independent variable
_____ **7.** The condition that is tested by the experimenter	**d.** experiment
_____ **8.** The condition being observed or measured in an experiment	**e.** control group

Use each of the terms below just once to complete the passage.

experimental results	experiment(s)	hypothesis	laws
scientific journals	theory	valid	verify

When **(9)** _____ are reported in **(10)** _____ ,

other scientists may try to **(11)** _____ the results by repeating the

(12) _____ . Usually when a(n) **(13)** _____ is supported by

data from several scientists, it is considered **(14)** _____ . Over time, a hypothesis that

is supported by many observations and experiments becomes a **(15)** _____ . Some

well-established facts of nature, such as gravity, are recognized as **(16)** _____ .

Chapter 1 **Biology: The Study of Life,** continued

In your textbook, read about kinds of research.

Complete the chart by checking the more correct column for each example.

Example	Quantitative Research	Qualitative Research
1. Numerical data		
2. Field study of hunting behavior		
3. Thermometer, balance scale, stopwatch		
4. Testable hypothesis		
5. Measurements from controlled laboratory experiments		
6. Purely observational data		
7. Binoculars, tape recorder, camera		

Complete each statement.

8. In order for scientific research to be universally understood, scientists report measurements in the

_____ , a modern form of the metric system.

9. This system of measurement is abbreviated _____ .

10. This system is a _____ system in which measurements are expressed in

multiples of _____·_____ or _____ of a basic unit.

In your textbook, read about science and society.

If the following statement is true, write _true_. If it is not, rewrite the italicized part to make it true.

11. Ideas about the value of knowledge gained through scientific research come from a society's *social*,

ethical, and *moral* concerns. _____

12. Pure science is scientific research carried out *primarily to solve a specific environmental problem.*

_____

13. *Technology* is the practical application of scientific research to improve human life and the world in

which we live. _____

14. A technological solution to a human problem can benefit humans but may also *cause a different, possibly*

serious, problem. _____

15. *Scientists* have the final say about how the results of scientific discoveries are applied.

_____

BioDigest

1 # What is biology?

In your textbook, read about characteristics of life.

Complete the following statements.

Biology is the study of **(1)** _____ and the **(2)** _____

among them. Biologists use a variety of **(3)** _____ methods to study the details

of life.

For each item in Column A, write the letter of the matching item in Column B.

Column A		Column B
_____	**4.** The basic unit is the cell.	**a.** development
_____	**5.** Maintenance of a stable internal environment	**b.** growth
_____	**6.** Reaction to a change in the environment	**c.** homeostasis
_____	**7.** Cell enlargement and division	**d.** organization
_____	**8.** Changes in an organism that take place over time	**e.** reproduction
_____	**9.** Transmission of heredity information from one generation to the next	**f.** response to stimulus

Using what you know about characteristics of life, determine if each of the following describes a living or nonliving thing.

_____ **10.** rust on a bucket   _____ **13.** lightning

_____ **11.** an apple on a tree   _____ **14.** a dinosaur fossil

_____ **12.** bacteria   _____ **15.** a wasp

BioDigest 1   **What is biology?,** *continued*

In your textbook, read about scientific methods.

Decide if each of the following statements is true. If it is not, rewrite the italicized part to make it true.

16. Scientific methods include observation, *forming a hypothesis*, and experiment.

_____

17. A statement that can be tested and presents a possible solution to a question is a *law*.

_____

18. In a controlled experiment, two groups are tested and all conditions except *two* are kept the same for both groups. _____

19. A condition that remains the same for both groups is called the *independent variable*.

_____

20. A condition that is changed by the experimenter in one group and not the other is called the

dependent variable. _____

21. A scientific experiment can be conducted *only in a laboratory*. _____

22. A theory is a *law* that has been confirmed by many experiments. _____

Read each of the following statements. If it is a testable hypothesis, write <u>yes</u>. If it is not a testable hypothesis, write <u>no</u>.

_____ **23.** If a person exercises, his or her pulse rate will increase.

_____ **24.** Cats make better pets than dogs.

_____ **25.** When fertilizer is added to soil, plants grow taller.

Identify each of the two italicized items as either an independent or a dependent variable.

26. The *number of red blood cells* in a mouse's blood at *different levels of iron* in its diet

_____

27. The *amount of starch formed* in a plant leaf for *different times* of exposure to light

_____

Chapter

2 **Principles of Ecology** *Section 2.1 Organisms and Their Environment*

In your textbook, read about what ecology is and about aspects of ecological study.

Use each of the terms below just once to complete the passage.

| ecology | biotic factors | nonliving | environments | atmosphere |
| humans | organisms | soil | biosphere | abiotic factors |

Living organisms in our world are connected to other **(1)** _____ in a

variety of ways. The branch of biology called **(2)** _____ is the scientific study of

interactions between organisms and their **(3)** _____ , including relationships

between living and **(4)** _____ things.

All living things on Earth can be found in the **(5)** _____ , the portion

of Earth that supports life. It extends from high in the **(6)** _____ to the bottom

of the oceans. Many different environments can be found in the biosphere. All living organisms found in

an environment are called **(7)** _____ . Nonliving parts of an environment are

called **(8)** _____ . For example, whales, trees, and **(9)** _____

are biotic factors. Ocean currents, temperature, and **(10)** _____ are abiotic factors.

In your textbook, read about levels of organization in ecology.

For each item in Column A, write the letter of the matching item in Column B.

Column A	Column B
_____ **11.** A group of organisms of one species that interbreed and live in the same place at the same time	**a.** community
	b. competition
_____ **12.** A collection of interacting populations	
	c. forest
_____ **13.** Interacting populations and abiotic factors in a community	
	d. population
_____ **14.** Increases when resources are scarce	
	e. ecosystem
_____ **15.** A terrestrial ecosystem	

In your textbook, read about organisms in ecosystems.

For each statement below, write true or false.

_____ **16.** A habitat is the role a species plays in a community.

_____ **17.** Habitats may change.

_____ **18.** A niche is the place where an organism lives its life.

_____ **19.** A habitat can include only one niche.

_____ **20.** A species' niche includes how the species meets its needs for food and shelter.

_____ **21.** The centipedes and worms that live under a certain log occupy the same habitat but have different niches.

_____ **22.** It is an advantage for two species to share the same niche.

_____ **23.** Competition between two species is reduced when the species have different niches.

Complete the table below by writing the kind of relationship described on the left.

Relationships Among Organisms	
Description of Relationship	**Kind of Relationship**
24. Organisms of different species live together in a close, permanent relationship.	
25. One species benefits and the other species is neither benefited nor harmed by the relationship.	
26. One species benefits from the relationship at the expense of the other species.	
27. Both species benefit from the relationship.	

Chapter 2 **Principles of Ecology,** *continued*

In your textbook, read about how organisms obtain energy and about matter and energy flow in ecosystems.

Answer the questions below. Use the diagram of a food web to answer questions 1–7.

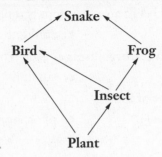

1. How many food chains make up the food web?

_____

2. Which organism is an herbivore?

_____

3. Which organism is an autotroph?

_____

4. Which organism is a third-order heterotroph? To what trophic level does that organism belong?

_____

5. Which organism is an omnivore?

_____

6. Which organisms belong to more than one food chain?

_____

7. Which organism belongs to more than one trophic level?

_____

8. What are decomposers? Where would decomposers appear in this food web?

_____

_____

9. What does a pyramid of energy show about the amount of energy available at different trophic levels of a food chain?

_____

_____

10. Why do different trophic levels have different amounts of energy?

_____

_____

Chapter 2

Principles of Ecology, *continued*

In your textbook, read about cycles in nature.

Circle the letter of the choice that best completes the statement or answers the question.

11. Energy that is lost at each trophic level of an ecosystem is replenished by
   a. heat.   **b.** nutrients.   **c.** sunlight.   **d.** organisms.

12. Besides energy, what moves through the organisms at each trophic level of an ecosystem?
   a. organisms   **b.** nutrients   **c.** sunlight   **d.** cycles

13. Evaporation and condensation a part of the
   a. carbon cycle.   **b.** nitrogen cycle.   **c.** phosphorus cycle.   **d.** water cycle.

14. Plants lose water to the air through
   a. condensation.   **b.** photosynthesis.   **c.** their roots.   **d.** evaporation.

15. Animals lose water when they
   a. breathe in.   **b.** urinate.   **c.** breathe out.   **d.** both b and c.

16. The major process by which water in the atmosphere is returned to the earth is
   a. precipitation.   **b.** evaporation.   **c.** photosynthesis.   **d.** decomposition.

17. Autotrophs and heterotrophs use carbon-containing molecules for energy and for
   a. photosynthesis.   **b.** growth.   **c.** decomposition.   **d.** both a and b.

18. What do plants use in photosynthesis to make carbon-containing molecules?
   a. carbon dioxide   **b.** carbohydrates   **c.** fertilizer   **d.** oxygen

19. Heterotrophs get carbon-containing molecules by
   a. making the molecules themselves.   **b.** feeding on other organisms.
   c. decaying.   **d.** growing.

20. When decomposers break down the carbon-containing molecules in dead organisms,
   a. the dead organisms are converted to coal.   **b.** oxygen is released.
   c. carbon dioxide is released.   **d.** carbon dioxide is converted to energy-rich carbon-containing molecules.

21. Fertilizers provide plants with
   a. nitrogen.   **b.** carbon.   **c.** water.   **d.** oxygen.

22. Which of the following convert(s) nitrogen from air into a form plants can use?
   a. bacteria   **b.** lightning   **c.** sunlight   **d.** both a and b

23. Plants use nitrogen to make
   a. carbohydrates.   **b.** nitrogen gas.   **c.** amino acids   **d.** both b and c.

24. An animal returns nitrogen to the environment when it
   a. breathes.   **b.** decomposes.   **c.** urinates.   **d.** both b and c.

25. Animals get phosphorus from
   a. the air.   **b.** eating plants.   **c.** water.   **d.** the soil.

26. Phosphorus in the soil comes from
   a. rocks.   **b.** decaying organisms.   **c.** the air.   **d.** both a and b.

Chapter 3 Communities and Biomes

In your textbook, read about living in a community.

If the statement is true, write "true" on the line. If it is not true, rewrite the italicized part to make it true.

1. The *interactions* of abiotic and biotic factors result in conditions that are suitable for some organisms but not for others. _____

2. Food availability and temperature can be *biotic factors* for a particular organism. _____

3. A limiting factor is any biotic or abiotic factor that *promotes* the existence, numbers, reproduction, or distribution of organisms. _____

4. At high elevations where the soil is thin, vegetation is limited to *large, deep-rooted* trees.

   _____

5. Factors that limit one population in a community may also have *an indirect* effect on another population.

   _____

6. *Tolerance* is the ability of an organism to withstand fluctuations in biotic and abiotic environmental factors. _____

7. A population of deer would become *larger* as conditions move away from optimal toward either extreme of the deer's range of tolerance. _____

8. Different species may have different ranges of tolerance. _____

In your textbook, read about succession: changes over time.

Use each of the terms below just once to complete the passage.

climax	primary	decades	succeed
pioneer	succession	species	slows down

The natural changes and **(9)** _____ replacements that take place in the communities

of ecosystems are know as **(10)** _____ . It can take **(11)** _____ or even centuries

for one community to **(12)** _____ , or replace, another. When new sites of land are formed, as

in a lava flow, the first organisms to colonize the new area are **(13)** _____ species. This colo-

nization is called **(14)** _____ succession. The species inhabiting the area gradually change.

Eventually, succession **(15)** _____ and the community becomes more stable. Finally, a mature

community that undergoes little or no change, called a **(16)** _____ community, develops.

For each item in Column A, write the letter of the matching item in Column B.

Column A	Column B
_____ **17.** Sequence of community changes where soil is formed, allowing small, weedy plants to inhabit the area	**a.** a severe drought
_____ **18.** Sequence of community changes occurring as a result of a natural disaster, such as a forest fire	**b.** primary succession
_____ **19.** A stable, mature community with little or no succession occurring	**c.** amount of plant growth
_____ **20.** An example of a biotic limiting factor affecting a community of organisms	**d.** secondary succession
_____ **21.** An example of an abiotic limiting factor affecting a community of organisms	**e.** climax community

The statements below describe the secondary succession that occurred within an area of Yellowstone National Park. Number the events in the order in which they occurred or will occur.

_____ **22.** Grasses, ferns, and pine seedlings inhabited the area.

_____ **23.** Annual wildflowers grew from the bare soil.

_____ **24.** A fire burned thousands of acres of land.

_____ **25.** A climax community of lodgepole pines developed.

_____ **26.** Bare soil covered the area.

Reinforcement and Study Guide

Section 3.2 Biomes

In your textbook, read about aquatic biomes: life in the water.

Complete each statement.

1. A large group of ecosystems sharing the same type of _____ is called a _____ .

2. Biomes located in bodies of _____ , such as oceans, lakes, and rivers, are called _____ .

3. The water in marine biomes is _____ .

4. Oceans contain the largest amount of _____ , or living material, of any biome on Earth. Yet, many of the organisms are so _____ that they cannot be seen without magnification.

5. The _____ is that part of marine biomes shallow enough to be penetrated by sunlight.

6. Deep-water regions of marine biomes receiving no sunlight make up the _____ .

Circle the letter of the response that best completes the statement.

7. Many rivers eventually flow into
   a. a lake.   **b.** a stream.   **c.** an ocean or a sea.   **d.** a swamp.

8. The body of water where fresh water from a river mixes with salt water is called
   a. an estuary.   **b.** a shoreline.   **c.** a sandbar.   **d.** a sea.

9. Organisms living in intertidal zones have structural adaptations that protect them from
   a. the dark.   **b.** sunlight.   **c.** wave action.   **d.** temperature.

10. Life is abundant in photic zones because
    a. there are no waves.   **b.** the water is cold.
    c. the water is clean.   **d.** there are many nutrients.

11. The greatest number of organisms living in the photic zone of a marine biome are
    a. dolphins.   **b.** plankton.   **c.** plants.   **d.** sharks.

12. Fewer organisms live at the bottom of a deep lake than at the top because of the lack of
    a. sunlight.   **b.** space.   **c.** niches.   **d.** bacteria.

In your textbook, read about terrestrial biomes.

Answer the following questions.

13. Which two abiotic factors generally determine the type of climax community that will develop in a particular part of the world?

_____

14. In which terrestrial biome is the ground permanently frozen?

_____

15. What are some adaptations that desert plants have developed?

_____

_____

_____

16. Describe the three layers of a tropical rain forest, including organisms that live in each layer.

_____

_____

_____

Write the name of each major terrestrial biome next to its description.

_____   **17.** Arid land with sparse, drought-resistant plants, minimal rainfall

_____   **18.** Largest terrestrial biome that supports small plants and grasses, but few trees

_____   **19.** Treeless land where ground remains frozen except for top few centimeters

_____   **20.** Warm, highly humid land that supports many species of organisms; extensive annual rainfall

_____   **21.** Land with coniferous forests, peat swamps, and long, harsh winters

_____   **22.** Land populated with broad-leaved hardwood trees that lose their leaves annually

Chapter 4 Population Biology

In your textbook, read about the principles of population growth.

Refer to Graphs A and B below. Answer the following questions.

Graph A

Graph B

1. What type of population growth is shown in Graph A? Explain this type of growth.

_____

_____

2. Which graph shows the most likely growth of a squirrel population living in a forest? _____

3. Which graph shows a population's growth under ideal conditions? _____

4. Why don't populations of organisms grow indefinitely?

_____

_____

_____

Use each of the terms below just once to complete the passage.

| grows | carrying capacity | below | births |
| above | under | deaths | exceed |

The number of organisms of one species that an environment can support is called its

(5) _____ . If the number of organisms in a population is **(6)** _____ the

environment's carrying capacity, births **(7)** _____ deaths and the population

(8) _____ . If the number of organisms rises **(9)** _____ the carrying capacity

of the environment, **(10)** _____ will exceed **(11)** _____ . This pattern will

continue until the population is once again at or **(12)** _____ the carrying capacity.

Circle the letter of the choice that best completes the statement.

13. The most important factor that determines population growth is the organism's
  a. social pattern.   **b.** carrying capacity.
  c. reproductive pattern. **d.** feeding pattern.

14. Organisms that follow a rapid life-history pattern
  a. have short life spans. **b.** have small bodies.
  c. reproduce early.   **d.** all of the above

15. Organisms that follow a slow life-history pattern
  a. have small bodies.   **b.** mature rapidly.
  c. reproduce slowly.   **d.** all of the above

16. A limiting factor that has an increasing effect as population size increases is
  a. temperature.   **b.** habitat disruption.
  c. drought.   **d.** competition.

In your textbook, read about how organism interactions limit population size.

Answer the following.

17. The snowshoe hare is a primary source of food for the Canadian lynx. Explain how the lynx population size changes when the hare population increases.

_____

_____

18. Explain how the change in the lynx population size affects the hare population.

_____

_____

19. What is the relationship between the lynx and the hare called?

_____

_____

20. When does competition decrease the size of a population?

_____

_____

21. What can cause an organism to exhibit stress, and what symptoms of stress can lead to a decrease in population size?

_____

_____

_____

In your textbook, read about demographic trends.

Determine if the statement is true. If it is not, rewrite the italicized part to make it true.

1. Looking at *past* population trends is a good way to predict the future of human populations.

_____

2. Demography is the study of population *health* characteristics.

_____

3. Worldwide human populations have *decreased* exponentially over the past few centuries.

_____

4. Humans are able to *increase* environmental effects on the human population through controlling disease, eliminating competing organisms, and increasing food production.

_____

5. To tell whether a population is *growing*, you must know the difference between the birthrate and the death rate.

_____

6. An *age structure* graph shows the proportions of a population at different age levels.

_____

7. A country with *high* doubling rate will double in size faster than one with a low doubling rate.

_____

8. Birthrates and death rates of countries around the world *are basically the same*.

_____

9. If a country has a high death rate, it *may also* have a high birthrate.

_____

10. If a country has a low death rate and a *high* birthrate, it will grow slowly, if at all.

_____

For each statement in Column A, write the letter of the item in Column B that completes the statement correctly.

Column A	Column B
_____ **13.** Population growth will change if the largest _____ of a population is in its post-reproductive years.	**a.** age structure
_____ **14.** The proportions of a population that are at different _____ make up its age structure.	**b.** stable
_____ **15.** If you know a population has a large group of individuals in their pre-reproductive years, you would predict that the population's growth will be _____ .	**c.** proportion
_____ **16.** If the proportions of a population at different age levels are fairly equal, the population will be _____ .	**d.** fertility
_____ **17.** The population growth of a country depends on its birthrate, death rate, and _____ rate.	**e.** rapid
_____ **18.** To make predictions about the growth of a population, demographers must know its _____ .	**f.** age levels

Complete each statement using the following choices: *world, emigration, immigration, a country's*

19. _____ is the movement of humans into a population.

20. _____ is the movement of humans from a population.

21. Immigration and emigration of people have no effect on total _____ population.

22. Immigration and emigration of people affect _____ population growth rates.

In your textbook, read about biological diversity.

Use the terms below just once to complete the passage. You will not use all the terms.

niches	variety	greater	space	species
biological diversity	equator	less	decrease	increase

(1) _____ refers to the (2) _____ of life in an

area. Another word for biological diversity is biodiversity. The simplest measure of biodiversity is the

number of (3) _____ that live in a certain area. The more species there are,

the (4) _____ is the biodiversity of the area. Biodiversity on land tends to

(5) _____ as you move toward the (6) _____ .

Biodiversity is greater on large islands than on small islands because large islands have more

(7) _____ and a greater variety of (8) _____ .

In your textbook, read about the importance of biodiversity.

For each statement below, write <u>true</u> or <u>false</u>.

_____ **9.** Biodiversity provides our world with beauty.

_____ **10.** The loss of a species from an ecosystem usually has no effect because of the presence of other species in the ecosystem.

_____ **11.** Biodiversity decreases the stability of ecosystems because more species are competing with each other.

_____ **12.** Increasing the biodiversity of an ecosystem may result in more niches.

_____ **13.** Diseases are more likely to spread in an ecosystem with high biodiversity than in an ecosystem with low biodiversity.

_____ **14.** A decrease in Earth's biodiversity may affect people's diets.

_____ **15.** Preserving diverse plant species may lead to the discovery of new drugs in the future.

Chapter 5 Biological Diversity and Conservation, *continued*

In your textbook, read about the loss of biodiversity.

For each item in Column A, write the letter of the matching item in Column B.

Column A	Column B
_____ **16.** The number of members of a species is so low that there is a possibility of extinction.	**a.** passenger pigeon
_____ **17.** This animal is an example of an endangered species.	**b.** threatened species
_____ **18.** The population of a species begins declining rapidly.	**c.** black rhinoceros
_____ **19.** This animal is an example of an extinct species.	**d.** African elephant
_____ **20.** All members of a species have died, so the species no longer exists.	**e.** extinct species
_____ **21.** This animal is an example of a threatened species.	**f.** endangered species

In your textbook, read about threats to biodiversity.

Complete the table by checking the most appropriate column for each statement.

Statement	Habitat Loss	Habitat Fragmentation	Habitat Degradation
22. Animals have no migratory route.			
23. A rain forest is burned.			
24. A highway divides a forest.			
25. Acid precipitation leaches nutrients from the soil.			
26. Detergents and other chemicals pollute bodies of water.			
27. Coral is mined for building materials.			
28. The reduction of the ozone layer causes more ultraviolet radiation to reach Earth's surface.			

Circle the letter of the choice that best completes the statement.

29. When species lose their habitats, they may
    a. lack food.
    b. lack shelter.
    c. be in danger of becoming extinct.
    d. all of the above.

30. Habitat fragmentation often leads to
    a. increased species diversity within an area.
    b. larger habitats for species.
    c. decreased species diversity within an area.
    d. an increased food supply for species.

31. Conditions where the boundaries of two ecosystems meet are called
    a. habitat fragmentation.   b. edge effect.   c. habitat loss.   d. canopy effect.

32. A great source of air pollution is
    a. volcanic eruptions.
    b. forest fires.
    c. burning fossil fuels.
    d. all of the above.

33. Acid precipitation
    a. may decrease biodiversity on land.
    b. has no effect on biodiversity.
    c. may increase biodiversity in water.
    d. both a and c.

34. The reduction of the ozone layer is caused by
    a. burning fossil fuels.   b. acid precipitation.   c. heavy metals.   d. CFCs.

35. Algal blooms in lakes
    a. are caused by acid precipitation.
    b. decrease the amount of oxygen in the lake when they decay.
    c. clog the gills of fish.
    d. both a and b.

36. When exotic species are introduced into an area, their populations may grow exponentially because the species
    a. are large.
    b. are predators.
    c. lack competitors and predators.
    d. are small.

In your textbook, read about strategies of conservation biology.

Answer the following questions.

1. What is conservation biology?

_____

_____

2. How does the U.S. Endangered Species Act protect biodiversity?

_____

_____

_____

3. How do nature preserves help protect biodiversity?

_____

_____

4. Why is it usually better to preserve one large area of land instead of a few smaller areas of land?

_____

_____

5. Why are habitat corridors used to connect different protected areas?

_____

6. What caused the steady decline of the black-footed ferret population in Wyoming?

_____

_____

7. What efforts were made to increase the size of the black-footed ferret population?

_____

_____

8. How are seed banks useful in protecting biodiversity?

_____

_____

_____

9. What are some problems of keeping endangered animals in captivity before reintroducing them to their original habitats?

_____

_____

2 Ecology

In your textbook, read about ecosystems.

For each statement below, write <u>true</u> or <u>false</u>.

_____ **1.** Organisms interact with the nonliving parts of their environments.

_____ **2.** Relationships between organisms are abiotic factors in ecosystems.

_____ **3.** In the carbon cycle, animals produce nutrients from carbon dioxide in the atmosphere.

_____ **4.** Commensalism is a relationship in which one species benefits while the other species is neither helped nor harmed.

_____ **5.** The temperature and precipitation in a certain land area influence the type of biome that is found there.

In your textbook, read about food for life.

Use the diagram on the right to answer questions 6–10.

6. Describe a food chain using organisms in the pyramid.

_____

_____

_____

_____

_____

7. Which organisms are carnivores?

_____

8. How many trophic levels are included in the pyramid?

_____

9. Which trophic level has the smallest biomass?

_____

10. How does the biomass of the autotrophs compare with the biomass of the herbivores?

_____

_____

BioDigest
2 **Ecology,** continued

In your textbook, read about population size.

Use the terms below to complete the passage. You will not use all the terms.

carrying capacity	species	maximum	limit	competition
linear growth	minimum	exceeds	food	exponential growth

A population is the number of organisms of one **(11)** _____ that live in a certain area. Under ideal conditions in which there are no factors that **(12)** _____ the size of a population, a population shows **(13)** _____ . However, in the environment, the sizes of populations are influenced by various limiting factors, such as the availability of **(14)** _____ , water, space, and other resources. As population size increases, **(15)** _____ for the resources increases. The **(16)** _____ size of a population that an environment can support is the environment's **(17)** _____ for that population. When a population **(18)** _____ the carrying capacity, individuals are unable to meet all their needs and die.

In your textbook, read about succession and biodiversity.

Number the steps of succession below in the order in which they occur.

_____ **19.** Shade from grasses and shrubs provides protection for tree saplings.

_____ **20.** Pioneer species and other small plants are unable to grow in the shade and die.

_____ **21.** Grasses and bushes appear.

_____ **22.** A plot of farmland is abandoned.

_____ **23.** Tree saplings grow and increase the amount of shade in the area.

_____ **24.** Pioneer species, such as dandelions, take root in the soil.

Answer the following questions.

25. What effect does succession have on the biodiversity of ecosystems?

_____

26. What human actions decrease the biodiversity of ecosystems?

_____

_____

_____

REINFORCEMENT AND STUDY GUIDE

Chapter 6 The Chemistry of Life

In your textbook, read about elements, atoms, and isotopes.

If the statement is true, write *true*. If it is not, rewrite the italicized part to make it true.

1. An element is a substance that *can be* broken down into simpler substances. _____

2. On Earth, *about 25* elements are essential to living organisms. _____

3. Only four elements—*carbon, hydrogen, oxygen, and nitrogen*—make up more than 96 percent of the mass of a human. _____

4. Each element is abbreviated by a one- or two-letter *formula*. _____

5. Trace elements, such as iron and magnesium, are present in living things in *very large* amounts.

_____

6. The properties of elements are determined by *the structures of their atoms.* _____

Label the parts of the atom. Use these choices:

energy level   electron   neutron   proton   nucleus

e^-

11. _____

7. _____

p^+
n^0

10. _____

8. _____

9. _____

Answer the following questions.

12. What is the maximum number of electrons in each of the following energy levels: first, second, third?

_____

_____

13. Boron has two isotopes, boron-10 and boron-11. Boron-10 has five protons and five neutrons. How many protons and neutrons does boron-11 have? Explain.

_____

_____

_____

_____

_____

Chapter 6 **The Chemistry of Life,** *continued*

Section 6.1 Atoms and Their Interactions

In your textbook, read about compounds and bonding, chemical reactions, and mixtures and solutions.

Write the type of substance described. Use these choices: compound, element.

_____ **14.** H_2O, a liquid that no longer resembles either hydrogen or oxygen gas

_____ **15.** A substance that can be broken down in a chemical reaction

_____ **16.** Carbon, the substance represented by the symbol C

Complete the table by checking the correct column for each description.

Statement	Ionic Bond(s)	Covalent Bond(s)
17. Found in the compound NaCl		
18. Increases the stability of atoms		
19. Results in the formation of a molecule		
20. Is formed when atoms share electrons		

Fill in the blanks with the correct number of molecules to balance the chemical equation. Then answer the questions.

$$C_6H_{12}O_6 + \underline{\quad} O_2 \longrightarrow \underline{\quad} CO_2 + \underline{\quad} H_2O$$

21. Why must chemical equations always balance?

_____

_____

22. Which number indicates the number of atoms of each element in a molecule of a substance.

_____

23. When is a mixture not a solution?

_____

_____

24. What is the difference between an acid and a base?

_____

_____

In your textbook, read about water and its importance.

For each statement below, write <u>true</u> or <u>false</u>.

_____ **1.** In a water molecule, electrons are shared equally between the hydrogen atoms and oxygen atom.

_____ **2.** The attraction of opposite charges between hydrogen and oxygen forms a weak oxygen bond.

_____ **3.** Because of its polarity, water can move from the roots of a plant up to its leaves.

_____ **4.** Water changes temperature easily.

_____ **5.** Unlike most substances, water expands when it freezes.

Circle the letter of the choice that best completes the statement.

6. All objects in motion have
  a. potential energy. **b.** heat energy. **c.** kinetic energy. **d.** random energy.

7. The first scientist to observe evidence of the random motion of molecules was
  a. Brown. **b.** Darwin. **c.** Mendel. **d.** Hooke.

8. The net movement of particles from an area of higher concentration to an area of lower concentration is called
  a. dynamic equilibrium. **b.** nonrandom movement.
  c. concentration gradient. **d.** diffusion.

9. Diffusion occurs because of
  a. nonrandom movement of particles. **b.** random movement of particles.
  c. a chemical reaction between particles. **d.** chemical energy.

10. When a few drops of colored corn syrup are added to a beaker of pure corn syrup, the color will
  a. move from low concentration to high concentration.
  b. form a polar bond.
  c. start to diffuse.
  d. remain on the bottom of the beaker.

11. Diffusion can be accelerated by
  a. decreasing the pressure. **b.** increasing the temperature.
  c. decreasing the movement of particles. **d.** increasing the dynamic equilibrium.

12. When materials pass into and out of a cell at equal rates, there is no net change in concentration inside the cell. The cell is in a state of
  a. dynamic equilibrium. **b.** metabolism. **c.** imbalance. **d.** inertia.

13. The difference in concentration of a substance across space is called
  a. dynamic equilibrium. **b.** concentration gradient.
  c. diffusion. **d.** Brownian movement.

Chapter 6 The Chemistry of Life, continued

Section 6.3 Life Substances

In your textbook, read about the role of carbon in organisms.

For each of the following statements about carbon, write <u>true</u> or <u>false</u>.

_____ **1.** Carbon atoms can bond together in straight chains, branched chains, or rings.

_____ **2.** Large molecules containing carbon atoms are called micromolecules.

_____ **3.** Polymers are formed by hydrolysis.

_____ **4.** Cells use carbohydrates for energy.

Write each item below under the correct heading.

sucrose	glucose	starch	$C_6H_{12}O_6$
cellulose	glycogen	fructose	$C_{12}H_{22}O_{11}$

Monosaccharide	Dissaccharide	Polysaccharide
5.	8.	10.
6.	9.	11.
7.		12.

Complete the table by checking the correct column for each description.

Description	Lipids	Proteins	Nucleic Acids
13. Made up of nucleotides			
14. Most consist of three fatty acids bonded to a glycerol molecule			
15. DNA and RNA			
16. Contain peptide bonds			
17. Produce proteins			
18. Commonly called fats and oils			
19. Made up of amino acids			
20. Used for long-term energy storage, insulation, and protective coatings			
21. Contain carbon, hydrogen, oxygen, and nitrogen			

Reinforcement and Study Guide

Section 7.1 The Discovery of Cells

In your textbook, read about the history of the cell theory.

For each statement in Column A, write the letter of the matching item in Column B.

Column A	Column B
_____ 1. The first scientist to describe living cells as seen through a simple microscope	**a.** Schleiden
_____ 2. Uses two or more glass lenses to magnify either living cells or prepared slides	**b.** compound light microscope
_____ 3. A scientist who observed that cork was composed of tiny, hollow boxes that he called cells	**c.** electron microscope
_____ 4. A scientist who concluded that all plants are composed of cells	**d.** Schwann
_____ 5. A scientist who concluded that all animals are composed of cells	**e.** Hooke
_____ 6. The microscope that allowed scientists to view molecules	**f.** van Leeuwenhoek

In your textbook, read about the two basic cell types.

Complete the table by checking the correct column for each statement.

Statement	Prokaryotes	Eukaryotes
7. Organisms that have cells lacking internal membrane-bound structures		
8. Do not have a nucleus		
9. Are either single-celled or made up of many cells		
10. Generally are single-celled organisms		
11. Organisms that have cells containing organelles		

Chapter 7 **A View of the Cell,** *continued*

In your textbook, read about maintaining a balance.

Use each of the terms below just once to complete the passage.

glucose	plasma membrane	homeostasis
organism	balance	selective permeability

Living cells maintain a **(1)** _____ by controlling materials that enter and leave. Without this

ability, the cell cannot maintain **(2)** _____ and will die. The cell must regulate internal con-

centrations of water, **(3)** _____ , and other nutrients and must eliminate waste products.

Homeostasis in a cell is maintained by the **(4)** _____ , which allows only certain

particles to pass through and keeps other particles out. This property of a membrane is known as

(5) _____ . It allows different cells to carry on different activities within the

same **(6)** _____ .

In your textbook, read about the structure of the plasma membrane.

For each statement below, write <u>true</u> or <u>false</u>.

_____ **7.** The structure and properties of the cell wall allow it to be selective and maintain homeostasis.

_____ **8.** The plasma membrane is a bilayer of lipid molecules with protein molecules embedded in it.

_____ **9.** A phospholipid molecule has a nonpolar, water-insoluble head attached to a long polar, soluble tail.

_____ **10.** The fluid mosaic model describes the plasma membrane as a structure that is liquid and very rigid.

_____ **11.** Eukaryotic plasma membranes can contain cholesterol, which tends to make the membrane more stable.

_____ **12.** Transport proteins span the cell membrane, allowing the selectively permeable membrane to regulate which molecules enter and leave a cell.

_____ **13.** Proteins at the inner surface of the plasma membrane attach the membrane to the cell's support structure, making the cell rigid.

In your textbook, read about cellular boundaries; nucleus and cell control; assembly, transport, and storage in the cell; and energy transformers.

Complete the table by writing the name of the cell part beside its structure/function. A cell part may be used more than once.

Structure/Function	Cell Part
1. A membrane-bound, fluid-filled sac	
2. Closely stacked, flattened membrane sacs	
3. The sites of protein synthesis	
4. A folded membrane that forms a network of interconnected compartments in the cytoplasm	
5. The clear fluid inside the cell	
6. Organelle that manages cell functions in eukaryotic cell	
7. Contains chlorophyll, a green pigment that traps energy from sunlight and gives plants their green color	
8. Digest excess or worn-out cell parts, food particles, and invading viruses or bacteria	
9. Small bumps located on the endoplasmic reticulum	
10. Provides temporary storage of food, enzymes, and waste products	
11. Firm, protective structure that gives the cell its shape in plants, fungi, most bacteria, and some protists	
12. Produce a usable form of energy for the cell	
13. Modifies proteins chemically, then repackages them	
14. Contains inner membranes arranged in stacks of membranous sacs called grana	
15. Plant organelles that store starches or lipids or that contain pigments	

Chapter 7 **A View of the Cell,** *continued*

In your textbook, read about structures for support and locomotion.

Determine if the statement is true. If it is not, rewrite the italicized part to make it true.

16. Cells have a support structure within the *cytoplasm* called the cytoskeleton.

_____

17. The *exoskeleton* is composed of thin, fibrous elements that form a framework for the cell.

_____

18. *Microtubules* of the cytoskeleton are thin, hollow cylinders made of protein.

_____

19. Cilia and flagella are cell surface structures that are adapted for *respiration*.

_____

20. *Flagella* are short, numerous, hairlike projections from the plasma membrane.

_____

21. Flagella are longer and *more* numerous than cilia.

_____

22. In *multicellular* organisms, cilia and flagella are the major means of locomotion.

_____

23. In *prokaryotic* cells, both cilia and flagella are composed of microtubules.

_____

Write titles for each of the generalized diagrams and then label the parts. Use these choices: plant cell, animal cell, plasma membrane, chloroplast, lysosome, large vacuole, cell wall.

24. _____   25. _____

26. _____

27. _____

28. _____

29. _____

30. _____

Cellular Transport and the Cell Cycle

In your textbook, read about osmosis: diffusion of water.

Complete the table by checking the correct column for each statement.

Statement	Isotonic Solution	Hypotonic Solution	Hypertonic Solution
1. Causes a cell to swell			
2. Doesn't change the shape of a cell			
3. Causes osmosis			
4. Causes a cell to shrink			

In your textbook, read about passive transport and active transport.

For each item in Column A, write the letter of the matching item in Column B.

Column A

_____ **5.** Transport protein that provides a tubelike opening in the plasma membrane through which particles can diffuse

_____ **6.** Is used during active transport but not passive transport

_____ **7.** Process by which a cell takes in material by forming a vacuole around it

_____ **8.** Particle movement from an area of higher concentration to an area of lower concentration

_____ **9.** Process by which a cell expels wastes from a vacuole

_____ **10.** A form of passive transport that uses transport proteins

_____ **11.** Particle movement from an area of lower concentration to an area of higher concentration

_____ **12.** Transport protein that changes shape when a particle binds with it

Column B

a. energy

b. facilitated diffusion

c. endocytosis

d. passive transport

e. active transport

f. exocytosis

g. carrier protein

h. channel protein

Chapter 8 — Cellular Transport and the Cell Cycle, *continued*

In your textbook, read about cell size limitations.

Determine if the statement is true. If it is not, rewrite the italicized part to make it true.

1. Most *living cells* are between 2 and 200 μm in diameter. _____

2. Diffusion of materials over long distance is *fast*. _____

3. If a cell doesn't have enough *DNA* to make all the proteins it needs, the cell cannot live.

_____

4. As a cell's size increases, its volume increases much *slower* than its surface area.

_____

5. If a cell's diameter doubled, the cell would require *two* times more nutrients and would have *two*

times more wastes to excrete. _____

In your textbook, read about cell reproduction.

Use each of the terms below just once to complete the passage.

nucleus	genetic material	chromosomes	packed
identical	chromatin	vanish	cell division

The process by which two cells are produced from one cell is called **(6)** _____ .

The two cells are **(7)** _____ to the original cell. Early biologists observed that just

before cell division, several short, stringy structures appeared in the **(8)** _____ .

These structures seemed to **(9)** _____ soon after cell division. These structures,

which contain DNA and became darkly colored when stained, are now called **(10)** _____ .

Scientists eventually learned that chromosomes carry **(11)** _____ , which

is copied and passed on from generation to generation. Chromosomes normally exist as

(12) _____ , long strands of DNA wrapped around proteins. However, before

a cell divides, the chromatin becomes tightly **(13)** _____ .

Chapter 8 **Cellular Transport and the Cell Cycle,** *continued*

Section 8.2 Cell Growth and Reproduction

In your textbook, read about the cell cycle and interphase.

Complete the table by checking the correct column for each statement.

Statement	Interphase	Mitosis
14. Cell growth occurs.		
15. Nuclear division occurs.		
16. Chromosomes are distributed equally to daughter cells.		
17. Protein production is high.		
18. Chromosomes are duplicated.		
19. DNA synthesis occurs.		
20. Cytoplasm divides immediately after this period.		
21. Mitochondria and other organelles are manufactured.		

In your textbook, read about the phases of mitosis.

Identify the following phases of mitosis. Use these choices: telophase, metaphase, anaphase, prophase. Then label the diagrams. Use these choices: sister chromatids, centromere, spindle fibers, centrioles.

22. _____ 23. _____ 24. _____ 25. _____

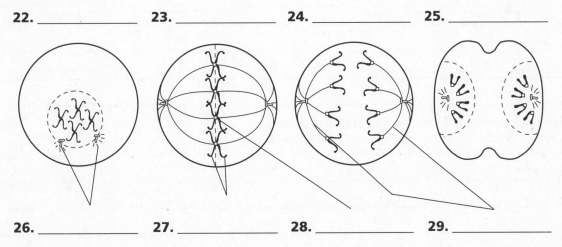

26. _____ 27. _____ 28. _____ 29. _____

Answer the question.

30. How does mitosis result in tissues and organs?

_____

_____

_____

Copyright © Glencoe/McGraw-Hill, a division of The McGraw-Hill Companies, Inc.

In your textbook, read about normal control of the cell cycle and cancer.

Answer the following questions.

1. In what ways do enzymes control the cell cycle?

_____

_____

2. What directs the production of these enzymes?

_____

_____

3. What can cause the cell cycle to become uncontrolled?

_____

_____

_____

_____

4. What can result when the cell cycle becomes uncontrolled?

_____

_____

5. What is the relationship between environmental factors and cancer?

_____

_____

_____

_____

6. What is a tumor? Describe the final stages of cancer.

_____

_____

_____

_____

7. Cancer is the second leading cause of death in the United States. What four types of cancer are the most prevalent?

_____

_____

Chapter
9 Energy in a Cell

Section 9.1 The Need for Energy

In your textbook, read about cell energy.

Use each of the terms below just once to complete the passage.

energy	phosphate	adenine	charged
ATP	chemical bonds	work	ribose

To do biological **(1)** _____ , cells require energy. A quick source

of energy that cells use is the molecule **(2)** _____ . The **(3)** _____

in this molecule is stored in its **(4)** _____ . ATP is composed of a(n)

(5) _____ molecule bonded to a(n) **(6)** _____ sugar.

Three **(7)** _____ molecules called **(8)** _____ groups

are attached to the sugar.

In your textbook, read about forming and breaking down ATP and the uses of cell energy.

Examine the diagram below. Then answer the questions.

9. How is energy stored and released by ATP?

_____

_____

_____

_____

10. How do cells use the energy released from ATP?

_____

_____

_____

_____

Chapter 9 **Energy in a Cell,** continued

Section 9.2 Photosynthesis: Trapping the Sun's Energy

In your textbook, read about trapping the sun's energy.

Determine if the statement is true. If it is not, rewrite the italicized part to make it true.

1. Photosynthesis is the process plants use to trap the sun's energy to make *glucose*.

_____

2. ATP molecules are made during the *light-independent* reactions of photosynthesis.

_____

3. *Carbon dioxide* gas is produced during photosynthesis.

_____

4. The light-dependent reactions of photosynthesis take place in the membranes of the thylakoid discs in *mitochondria*.

_____

5. The thylakoid membranes contain chlorophyll and other pigments that *absorb* sunlight.

_____

In your textbook, read about the light-dependent reactions of photosynthesis.

Number the following steps of the light-dependent reactions in the order in which they occur.

_____ **6.** The energy lost by electrons as they pass through the electron transport chain is used to make ATP.

_____ **7.** The electrons pass from the chlorophyll to an electron transport chain.

_____ **8.** Sunlight strikes the chlorophyll molecules in the thylakoid membranes.

_____ **9.** NADP+ molecules change to NADPH as they carry the electrons to the stroma of the chloroplast.

_____ **10.** The sunlight's energy is transferred to the chlorophyll's electrons.

_____ **11.** The electrons are passed down a second electron transport chain.

Answer the following questions.

12. How are the electrons that are lost by the chlorophyll molecules replaced?

_____

_____

13. How do plants produce oxygen during photosynthesis?

_____

_____

Chapter 9 Energy in a Cell, *continued*

Section 9.2 Photosynthesis: Trapping the Sun's Energy

In your textbook, read about the light-independent reactions.

Circle the letter of the choice that best completes the statement or answers the question.

14. The Calvin cycle includes
    a. the light-dependent reactions.
    b. an electron transport chain.
    c. the light-independent reactions.
    d. photolysis.

15. The Calvin cycle takes place in the
    a. mitochondria.
    b. stroma.
    c. nucleus.
    c. thylakoid membrane.

16. What product of the light-dependent reactions is used in the Calvin cycle?
    a. oxygen
    b. carbon dioxide
    c. NADPH
    d. chlorophyll

17. What gas is used in the first step of the Calvin cycle?
    a. oxygen
    b. carbon dioxide
    c. hydrogen
    d. water

18. A carbon atom from carbon dioxide is used to change the five-carbon sugar RuBP into
    a. ATP.
    b. two molecules.
    c. PGA.
    d. a six-carbon sugar.

19. How many molecules of the three-carbon sugar PGA are formed?
    a. two
    b. one
    c. six
    d. three

20. ATP, NADPH, and hydrogen ions are used to convert PGA into
    a. PGAL.
    b. glucose.
    c. RuBP.
    d. carbon dioxide.

21. How many rounds of the Calvin cycle are needed to form one glucose molecule?
    a. one
    b. six
    c. two
    d. three

22. What two molecules leave the Calvin cycle and are combined to form glucose?
    a. RuBP
    b. PGA
    c. PGAL
    d. CO_2

23. Which molecule from the Calvin cycle is used to replenish the five-carbon sugar, RuBP, which is used at the beginning of the cycle?
    a. NADP
    b. CO_2
    c. PGA
    d. PGAL

Chapter 9

Energy in a Cell, *continued*

Section 9.3 Getting Energy to Make ATP

In your textbook, read about the cellular respiration and fermentation.

Fill in the names of the molecules to complete the glycolysis reaction. Use these choices: 2PGAL, 4ATP, glucose, 2ADP, 2 pyruvic acid, 2NADH + 2H⁺. Then answer the questions.

Glycolysis

2ATP   **2.** _____ $4ADP + 4P_i$ **4.** _____

1. _____   **3.** _____   **5.** _____

$2NAD^+$ **6.** _____

7. What happens to pyruvic acid before entering the citric acid cycle?

_____

8. What happens to the electrons carried by the NADH and $FADH_2$ molecules produced during the citric acid cycle?

_____

_____

9. During which stages of cellular respiration are ATP molecules formed?

_____

10. Why is oxygen necessary for cellular respiration?

_____

11. How is fermentation different from cellular respiration?

_____

_____

In your textbook, read about comparing photosynthesis and cellular respiration.

Answer the following question.

12. Describe two ways in which cellular respiration is the opposite of photosynthesis.

_____

_____

_____

_____

BioDigest

3 **The Life of a Cell**

In your textbook, read about the chemistry of life.

Label the diagram below, using these choices:

atom   electron   molecule   neutron   nucleus   proton

1. _____

2. _____

3. _____

4. _____

5. _____

6. _____

In your textbook, read about eukaryotic cells, prokaryotic cells, and organelles.

Complete each statement.

7. Every cell is surrounded by a plasma _____ .

8. _____ contain membrane-bound structures called organelles within the cell.

9. Organisms having cells without internal membrane-bound structures are called _____ .

10. The plasma membrane is composed of a _____ with embedded proteins.

11. The _____ controls cell functions.

12. Ribosomes are organelles found in the cytoplasm that produce _____ .

13. The _____ and Golgi apparatus transport and modify proteins.

14. Plant cells contain _____ that capture the sun's light energy so that it can be transformed into usable chemical energy.

15. A network of microfilaments and microtubules attached to the cell membrane give the cell _____ .

16. _____ are long projections from the surface of the plasma membrane and move in a whiplike fashion to propel a cell.

In your textbook, read about diffusion and osmosis.

Answer the following questions.

17. What is diffusion? _____
_____

18. What is osmosis? _____
_____

19. What is active transport? _____
_____

In your textbook, read about mitosis.

For each item in Column A, write the letter of the matching item in Column B.

Column A		Column B

_____ **20.** Duplicated chromosomes condense and mitotic spindles form on the two opposite ends of the cell.

_____ **21.** Chromosomes slowly separate to opposite ends of cells.

_____ **22.** Chromosomes uncoil, spindle breaks down, and nuclear envelope forms around each set of chromosomes.

_____ **23.** Cells experience a period of intense metabolic activity prior to mitosis.

_____ **24.** Chromosomes line up in center of cell.

a. anaphase

b. interphase

c. metaphase

d. prophase

e. telophase

In your textbook, read about energy in a cell.

Decide if each of the following statements is true. If it is not, rewrite the italicized part to make it true.

_____ **25.** Adenosine triphosphate (ATP) is the most commonly used source of *protein* in a cell.

_____ **26.** *Light-dependent* reactions convert energy into starch through the Calvin cycle.

_____ **27.** *Mitochondria* convert food energy to ATP through a series of chemical reactions.

_____ **28.** Glycolysis produces a net gain of two ATP for *every two molecules* of glucose.

Chapter 10 Mendel and Meiosis

Section 10.1 Mendel's Laws of Heredity

In your textbook, read why Mendel succeeded.

Complete each statement.

1. Mendel was the first person to succeed in predicting how traits are _____ from generation to generation.

2. Mendel used _____ plants in his experiments.

3. In peas, both male and female sex cells—_____—are in the same flower.

4. _____ occurs when the male gamete fuses with the female gamete.

5. Mendel used the process called _____ when he wanted to breed one plant with another.

6. Mendel carefully _____ his experiments and the peas he used.

7. Mendel studied only one _____ at a time and analyzed his data mathematically.

In your textbook, read about Mendel's monohybrid crosses.

Refer to the table of pea-plant traits on the right. Then complete the table on the left by filling in the missing information for each cross. The first one is done for you.

Parent Plants	F$_1$ generation	
	Offspring	Appearance
8. round × wrinkled $RR \times rr$	Rr	round
9. yellow × green $YY \times yy$	a.	b.
10. axial × terminal $AA \times$ _____	Aa	a.
11. tall × short _____ × _____	Tt	a.
12. inflated × constricted _____ × ii	a.	b.

Pea-Plant Traits		
Trait	**Dominant**	**Recessive**
seed shape	round (R)	wrinkled (r)
seed color	yellow (Y)	green (y)
flower position	axial (A)	terminal (a)
plant height	tall (T)	short (t)
pod shape	inflated (I)	constricted (i)

In your textbook, read about phenotypes and genotypes and Mendel's dihybrid crosses.

If the statement is true, write true. If it is not, rewrite the underlined part to make it true.

13. A pea plant with the genotype *TT* has the same phenotype as a pea plant with genotype *tt*. _____

14. When Mendel crossed true-breeding pea plants that had round yellow seeds with true-breeding pea plants that had wrinkled green seeds, some of the offspring had round yellow seeds because round and yellow were the dominant forms of the traits. _____

15. When Mendel allowed heterozygous F_1 plants that had round yellow seed to self-pollinate, he found that some of the F_2 plants had wrinkled green seeds. _____

16. The law of independent assortment states that genes for different traits are inherited independently of each other. _____

In your textbook, read about Punnett squares and probability.

The Punnett square below is for a dihybrid cross between pea plants that are heterozygous for seed shape (*Rr*) and seed color (*Yy*). Complete the Punnett square by recording the expected genotypes of the offspring. Then answer the questions.

	RY	*Ry*	*rY*	*ry*
RY				
Ry				
rY				
ry				

17. Use the chart on the previous page to determine the phenotypes of the offspring. Record the phenotypes below the genotypes in the Punnett square. Is an offspring produced by the cross more likely to have wrinkled seeds or round seeds? _____

18. What is the probability that an offspring will have wrinkled yellow seeds? _____

Chapter 10 Mendel and Meiosis, *continued*

In your textbook, read about genes, chromosomes, and numbers.

Examine the table. Then answer the questions.

Chromosome Numbers of Some Common Organisms

Organism	Body Cell (2n)	Gamete (n)
Human	46	23
Garden pea	14	7
Fruit fly	8	4
Tomato	24	12
Dog	78	39
Chimpanzee	48	24
Leopard frog	26	13
Corn	20	10

1. What is the diploid number of chromosomes in corn?

_____

2. What is the haploid number of chromosomes in corn?

_____

3. Is the chromosome number related to the complexity of the organism?

_____

4. How many pairs of chromosomes do humans have?

_____

5. What process maintains a constant number of chromosomes within a species?

_____

In your textbook, read about the phases of meiosis.

Label the diagrams below. Use these choices: Metaphase I, Metaphase II, Interphase, Telophase I, Telophase II, Anaphase I, Anaphase II, Prophase I, Prophase II.

6. _____   7. _____   8. _____   9. _____   10. _____

11. _____   12. _____   13. _____   14. _____

The following statements describe interphase and and meiosis I. Identify each phase. Then place them in sequential order using the numbers 1 through 5. Use 1 for the phase that occurs first and 5 for the phase that occurs last.

Statement	Name of Phase	Sequence
15. Homologous chromosomes line up at the equator in pairs.	a.	b.
16. The cell replicates its chromosomes.	a.	b.
17. Homologous chromosomes separate and move to opposite ends of the cell.	a.	b.
18. The spindle forms, and DNA coils up and homologous chromosomes come together in a tetrad; crossing over may occur.	a.	b.
19. Events occur in the reverse order from the events of prophase I. Each cell has only half the genetic information; however, another cell division is needed because each chromosome is still doubled.	a.	b.

In your textbook, read about how meiosis provides for genetic variation and about mistakes in meiosis.

For each statement below, write <u>true</u> or <u>false</u>.

_____ **20.** Reassortment of chromosomes can occur during meiosis by crossing over of nonsister chromatids or by independent segregation of homologous chromosomes.

_____ **21.** Genetic recombination is a major source of variation among organisms.

_____ **22.** The random segregation of chromosomes during meiosis explains Mendel's observation that genes for different traits are inherited independently of each other.

_____ **23.** Nondisjunction always results in a zygote with an extra chromosome.

_____ **24.** Down syndrome is a result of polyploidy.

_____ **25.** Mistakes in meiosis can occasionally be beneficial.

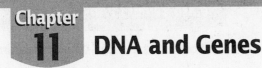

Chapter 11 · DNA and Genes

Section 11.1 DNA: The Molecule of Heredity

In your textbook, read about what DNA is and the replication of DNA.

Label the diagram. Use these choices: nucleotide, deoxyribose, phosphate group, nitrogenous base, hydrogen bonds, base pair.

1. _____

2. _____

3. _____

4. _____

5. _____

6. _____

Complete each statement.

7. _____ , guanine (G), cytosine (C), and thymine (T) are the four _____ in DNA.

8. In DNA, _____ always forms hydrogen bonds with guanine (G).

9. The sequence of _____ carries the genetic information of an organism.

10. The process of _____ produces a new copy of an organism's genetic information, which is passed on to a new cell.

11. The double-coiled shape of DNA is called a _____ .

Chapter 11 **DNA and Genes,** *continued*

In your textbook, read about genes and proteins and RNA.

Complete the chart on the three chemical differences between DNA and RNA.

Structure	DNA	RNA
1. strand of nucleotides	a.	b.
2. sugar	a.	b.
3. nitrogenous base	a.	b.

In your textbook, read about the genetic code.

Complete each statement.

4. Proteins are made up of _____ .

5. There are twenty different types of _____ .

6. The message of the DNA code is information for building _____ .

7. Each set of three nitrogenous bases that codes for an amino acid is known as a

_____ .

8. The amino acid _____ is represented by the mRNA codon ACA.

9. _____ and _____ are mRNA codons for phenylalanine.

10. There can be more than one _____ for the same amino acid.

11. For any one codon, there can be only one _____ .

12. The genetic code is said to be universal because a codon represents the same

_____ in almost all organisms.

13. _____ , _____ , and _____ are stop codons.

14. _____ and _____ are amino acids that are each

represented by only one codon.

Copyright © Glencoe/McGraw-Hill, a division of The McGraw-Hill Companies, Inc.

Chapter 11 **DNA and Genes,** *continued*

Section 11.2 From DNA to Protein

In your textbook, read about transcription from DNA to mRNA.

Complete each statement.

15. Proteins are made in the cytoplasm of a cell, whereas DNA is found only in the

_____ .

16. The process of making RNA from DNA is called _____ .

17. The process of transcription is similar to the process of DNA _____ .

18. _____ carries information from the DNA in the nucleus out into the cyto-

plasm of the cell.

19. mRNA carries the information for making proteins to the _____ .

In your textbook, read about translation from mRNA to protein.

Label the diagram. Use these choices: transfer RNA (tRNA), amino acid, amino acid chain, codon, anticodon, messenger RNA (mRNA), ribosome.

20. _____

21. _____

22. _____

23. _____

24. _____

25. _____

26. _____

Chapter 11 **DNA and Genes,** *continued*

In your textbook, read about mutation: a change in DNA.

Circle the letter of the choice that best completes the statement.

1. A mutation is any mistake or change in the
    a. cell.   **b.** DNA sequence.   **c.** ribosomes.   **d.** nucleus.

2. A point mutation is a change in
    a. several bases in mRNA.   **b.** several bases in tRNA.
    c. a single base pair in DNA.   **d.** several base pairs in DNA.

3. A mutation in which a single base is added to or deleted from DNA is called
    a. a frame shift mutation.   **b.** a point mutation.   **c.** translocation.   **d.** nondisjunction.

4. Chromosomal mutations are especially common in
    a. humans.   **b.** animals.   **c.** bacteria.   **d.** plants.

5. Few chromosome mutations are passed on to the next generation because
    a. the zygote usually dies.
    b. the mature organism is sterile.
    c. the mature organism is often incapable of producing offspring.
    d. all of the above.

6. When part of one chromosome breaks off and is added to a different chromosome, the result is a(n)
    a. translocation.   **b.** insertion.   **c.** inversion.   **d.** deletion.

7. Many chromosome mutations result when chromosomes fail to separate properly during
    a. mitosis.   **b.** meiosis.   **c.** crossing over.   **d.** linkage.

8. The failure of homologous chromosomes to separate properly is called
    a. translocation.   **b.** disjunction.   **c.** nondisjunction.   **d.** deletion.

9. Mutations that occur at random are called
    a. spontaneous mutations.   **b.** nonspontaneous mutations.
    c. nonrandom mutations.   **d.** environmental mutations.

10. An agent that can cause a change in DNA is called a(n)
    a. zygote.   **b.** inversion.
    c. mutagen.   **d.** mutation.

11. Mutations in body cells can sometimes result in
    a. new species.   **b.** cancer.
    c. sterile offspring.   **d.** hybrids.

Chapter 12 Patterns of Heredity and Human Genetics

In your textbook, read about making a pedigree.

Examine the pedigree to the right. Then answer the following questions.

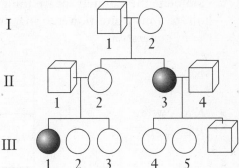

1. Is the trait being studied in the pedigree recessive or dominant? How do you know?

_____

_____

_____

_____

2. Are II-1 and II-2 carriers of the trait? How do you know?

_____

_____

3. What is the probability that II-1 and II-2 will produce an individual with the trait being studied? Draw a Punnett square to show your work.

_____

_____

_____

_____

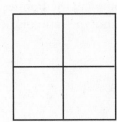

4. What is the likely genotype of II-4 for the trait being studied in the pedigree?

_____

In your textbook, read about simple recessive heredity and simple dominant heredity.

For each item in Column A, write the letter of the matching item from Column B.

Column A	Column B
_____ **5.** Recessive disorder that results from the absence of an enzyme required to break lipids down	**a.** cystic fibrosis
_____ **6.** Lethal genetic disorder caused by a dominant allele	**b.** simple dominant traits
_____ **7.** Most common genetic disorder among white Americans	**c.** Tay-Sachs disease
_____ **8.** Recessive disorder that results from the absence of an enzyme that converts one amino acid into another one	**d.** Huntington's disease
_____ **9.** Widow's peak and hitchhiker's thumb	**e.** phenylketonuria

Chapter 12 **Patterns of Heredity and Human Genetics,** *continued*

Section 12.2 When Heredity Follows Different Rules

In your textbook, read about complex patterns of inheritance.

Answer the following questions.

1. Complete the Punnett square for a cross between a homozygous red-flowered snapdragon (RR) and a homozygous white-flowered snapdragon ($R'R'$). Give the genotype and phenotype of the offspring in the F_1 generation.

Key
RR - red
$R'R'$ - white F_1
RR' - pink

genotype: _____

phenotype: _____

2. When traits are inherited in an incomplete dominance pattern, what is true of the phenotype of the heterozygotes?

_____

3. Complete the Punnett square for a cross between two pink-flowered (RR') F_1 plants. Give the phenotype ratio of the offspring in the F_2 generation.

F_2

phenotype ratio: _____

4. In what type of inheritance are both alleles expressed equally?

_____

5. Complete the Punnett square for a cross between a black chicken (BB) and a white chicken (WW). Give the phenotype of the offspring in the F_1 generation.

Key
BB - black
WW - white F_1
BW - checkered

phenotype: _____

Chapter 12 Patterns of Heredity and Human Genetics, *continued*

Section 12.2 When Heredity Follows Different Rules

For each statement below, write true or false.

_____ **6.** Traits controlled by more than two alleles are said to have multiple alleles.

_____ **7.** Multiple alleles can be studied only in individuals.

_____ **8.** In humans, there are 23 pairs of matching homologous chromosomes called autosomes.

_____ **9.** Two chromosomes called the sex chromosomes determine the sex of an individual.

_____ **10.** The sex chromosomes of a human male are XX, and the sex chromosomes of a human female are XY.

_____ **11.** Traits controlled by genes located on sex chromosomes are called sex-linked traits.

_____ **12.** The first known example of sex-linked inheritance was discovered in pea plants.

In your textbook, read about environmental influences.

Answer the following questions.

13. What characteristics of an organism can affect gene function?

_____

_____

14. Do the internal environments of males and females differ? Explain.

_____

_____

15. What are some environmental factors that can influence gene expression?

_____

_____

16. Give two examples of how an environmental factor can affect the expression of a phenotype.

_____

_____

In your textbook, read about multiple alleles in humans.

Complete the table by filling in the missing information.

Genotypes	Human Blood Groups Surface Molecules	Phenotypes
1. $I^A I^A$ or $I^A i$	A	
2. $I^B I^B$ or $I^B i$		B
3.	A and B	AB
4.	none	

Complete each statement.

5. Blood groups are a classic example of _____ inheritance in humans.

6. The alleles _____ are always both expressed.

7. The alleles I^A and I^B are _____ , meaning they are always both expressed.

8. I^A and I^B are dominant to _____ .

9. Blood typing is necessary before a person can receive a _____ .

10. A child who inherits I^A from his mother and I^B from his father will have type _____ blood.

11. A child whose parents both have type O blood will have type _____ blood.

12. If a woman with blood type A has a baby with blood type AB, a man with blood type O _____ be the father.

13. Blood tests _____ be used to prove that a certain man is the father of a child.

Chapter 13 Genetic Technology

In your textbook, read about selective breeding and determining genotypes.

Complete each statement.

1. Organisms that are homozygous dominant and those that are _____ for a trait controlled by Mendelian inheritance have the same phenotype.

2. A _____ determines whether an organism is heterozygous or homozygous dominant for a trait.

3. Usually the parent with the known genotype is _____ for the trait in question.

4. When two cultivars are crossed, their offspring will be _____ .

Answer the following.

5. A breeder performs a testcross to determine whether an Alaskan malamute is homozygous dominant (*DD*) or heterozygous (*Dd*) for a recessive dwarf allele. Half the offspring appear dwarf. What is the genotype of the unknown dog? Complete the Punnett square to verify your answer.

_____

6. What results would be expected if the unknown dog was homozygous dominant (*DD*)? Complete the Punnett square to verify your answer.

_____

_____

In your textbook, read about gene engineering.

For each item in Column A, write the letter of the matching item in Column B.

Column A	Column B
_____ 1. Bacterial proteins that have the ability to cut both strands of the DNA molecule at certain points	**a.** recombinant DNA
_____ 2. Contain foreign DNA	**b.** vector
_____ 3. Is made by connecting segments of DNA from different sources	**c.** restriction enzymes
_____ 4. General term for a carrier used to transfer a foreign DNA fragment into a host cell	**d.** plasmid
_____ 5. A small ring of DNA found in a bacterial cell	**e.** transgenic organisms
_____ 6. The procedure for cleaving DNA from an organism into small segments, and inserting the segments into another organism	**f.** genetic engineering or recombinant DNA technology

Complete the table by checking the correct column for each vector.

Vectors	Mechanical	Biological
7. Viruses		
8. Micropipette		
9. Metal bullets		
10. Plasmids		

Chapter 13 Genetic Technology, *continued*

Section 13.2 Recombinant DNA Technology

In your textbook, read about applications of DNA technology.

Complete the table by checking the correct column for each statement.

Statement	Industry	Medicine	Agriculture
11. Recombinant DNA produces molecules to treat diseases.			
12. Engineered bacteria produce indigo dye.			
13. Higher yields of oil from canola			
14. Corrected human hemophilia gene			
15. Herbicide resistance			
16. Production of cheese and more effective laundry detergents			
17. Production of growth hormone to treat pituitary dwarfism and insulin to treat diabetes			
18. Development of nonallergenic foods			
19. Produce *Bt* toxin from bacterial gene			
20. Recombinant DNA used to make human antibodies			
21. Studying transgenic animals provides understanding of the role that specific genes play			
22. Transgenic sheep produce a protein that helps lungs inflate			
23. Recombinant DNA techniques that increase enzyme activity, stability, and specificity.			

Reinforcement and Study Guide

Section 13.3 **The Human Genome**

In your textbook, read about mapping and sequencing the human genome and applications of the Human Genome Project.

Determine if the statement is true. If it is not, rewrite the italicized part to make it true.

1. The human genome consists of approximately *1000* genes located on 46 chromosomes.

_____

2. Scientists *have* determined the exact chromosomal location of all genes.

_____

3. The genetic map that shows the location of genes on a chromosome is called a *pedigree map*.

_____

4. Instead of examining actual offspring, scientists examine *egg* cells to create linkage maps.

_____

5. Gene therapy is being performed on patients suffering from *sickle-cell anemia*.

_____

6. *Electrolysis* can be used to separate DNA fragments.

_____

Answer the following questions.

7. What is the Human Genome Project?

_____

_____

8. Why is mapping by linkage data extremely inefficient in humans?

_____

_____

_____

9. What are the three areas of current research that utilize chromosome maps?

_____

_____

10. Why is DNA fingerprinting reliable?

_____

_____

_____

REINFORCEMENT AND STUDY GUIDE

BioDigest
4 Genetics

In your textbook, read about simple Mendelian inheritance and meiosis.

Complete each statement.

1. A trait is _____ if only one allele is needed for that trait to be expressed. If both alleles are needed for the trait to be expressed, the trait is _____ .

2. When a *TT* tall pea plant is crossed with a *tt* short pea plant, there is a 100% probability that all off-spring will be _____ and have the genotype _____ .

3. Unlike mitosis, _____ produces cells that contain only one copied chromosome of each _____ .

4. _____ and the rearrangement of alleles during _____ provide mechanisms for genetic variability.

In your textbook, read about producing physical traits and complex inheritance patterns.

Predict the outcome of the following crosses. Use Punnett squares to support your answers.

5. Homozygous short × Homozygous short

_____

6. Heterozygous for purple flowers × Heterozygous for purple flowers

_____

_____

7. Heterozygous pink snapdragon × Heterozygous pink snapdragon

_____

_____

_____

_____

Sequence the steps in protein synthesis from 1 to 4.

_____ **8.** Amino acids bond together to form a protein.

_____ **9.** Sequence of bases in DNA is copied into mRNA.

_____ **10.** tRNA molecules bring appropriate amino acids to the mRNA on the ribosome.

_____ **11.** mRNA leaves the cell nucleus.

For each item in Column A, write the letter of the matching item in Column B.

Column A	Column B
_____ **12.** Results in an mRNA copy of DNA	**a.** incomplete dominance
_____ **13.** Sequence of three bases in mRNA	**b.** X-linked trait
_____ **14.** Site of translation	**c.** ribosome
_____ **15.** Governed by several genes	**d.** transcription
_____ **16.** Heterozygote has an intermediate phenotype.	**e.** translation
_____ **17.** Double stranded molecule that stores and transmits genetic information	**f.** codon
_____ **18.** More likely to appear in males than in females	**g.** polygenic inheritance
_____ **19.** Results in a sequence of amino acids	**h.** DNA

In your textbook, read about recombinant DNA technology and gene therapy.

Sequence the steps to making recombinant DNA from 1 to 5.

_____ **20.** The plasmid becomes part of a host cell's chromosome.

_____ **21.** A DNA fragment is inserted into a plasmid.

_____ **22.** The DNA fragment replicates during cell division.

_____ **23.** The plasmid enters a host bacterial cell.

_____ **24.** A host cell produces a protein that it would not have produced naturally.

Answer the following questions.

25. What is gene therapy?

_____

_____

26. What are clones?

_____

_____

27. What is a vector? Give two examples of vectors.

_____

_____

_____

In your textbook, read about the early history of Earth.

For each statement below, write <u>true</u> or <u>false</u>.

_____ **1.** Earth is thought to have formed about 4.6 billion years ago.

_____ **2.** The conditions on primitive Earth were very suitable for life.

_____ **3.** Geological events on Earth set up conditions that would play a major role in the evolution of life on Earth.

_____ **4.** By the end of the Mesozoic, the continents took on their modern shape.

_____ **5.** The first organisms appeared on land between 3.9 and 3.4 billion years ago.

In your textbook, read about a history in the rocks.

For each statement in Column A, write the letter of the matching item in Column B.

Column A	Column B
_____ **6.** A footprint, trail, or burrow, providing evidence of animal activity	**a.** petrified fossil
_____ **7.** A fossil embedded in tree sap, valuable because the organism is preserved intact	**b.** imprint
_____ **8.** An exact stone copy of an original organism, the hard parts of which have been penetrated and replaced by minerals	**c.** trace fossil
_____ **9.** Any evidence of an organism that lived long ago	**d.** cast
_____ **10.** The fossil of a thin object, such as a leaf or feather, that falls into sediments and leaves an outline when the sediments hardened	**e.** amber-preserved
_____ **11.** An empty space left in rock, showing the exact shape of the organism that was buried and decayed there	**f.** fossil
_____ **12.** An object formed when a mold is filled in by minerals from the surrounding rock	**g.** mold

In your textbook, read about the age of a fossil.

Answer the following questions.

13. Explain how relative dating works.

_____

_____

_____

14. What is the limitation of relative dating?

_____

15. What dating technique is often used by paleontologists to determine the specific age of a fossil?

_____

16. How do scientists use this dating technique to determine the ages of rocks or fossils?

_____

_____

In your textbook, read about a trip through geologic time.

Complete the table by checking the correct column for each statement.

Statement	Era			
	Pre-Cambrian	Paleozoic	Mesozoic	Cenozoic
17. The first photosynthetic bacteria form dome-shaped structures called stromatolites.				
18. Primates evolve and diversify.				
19. Divided into three periods: Triassic, Jurassic, and Cretaceous				
20. An explosion of life, characterized by the appearance of many types of invertebrates and plant phyla				
21. Mammals appear.				
22. Dinosaurs roam Earth, and the ancestors of modern birds evolve.				
23. Flowering plants appear.				
24. Amphibians and reptiles appear.				

Chapter 14 The History of Life, *continued*

In your textbook, read about origins: the early ideas.

Use each of the terms below just once to complete the passage.

microorganisms	vital force	Louis Pasteur	biogenesis
nonliving matter	S-shaped	disproved	Francesco Redi
organisms	broth	microscope	spontaneous generation
spontaneously	air		

Early scientists believed that life arose from **(1)** _____ through a process they called

(2) _____ . In 1668, the Italian physician **(3)** _____ conducted

an experiment with flies that **(4)** _____ this idea. At about the same time, biologists

began to use an important new research tool, the **(5)** _____ . They soon discovered the

vast world of **(6)** _____ . The number and diversity of these organisms was so great that

scientists were led to believe once again that these organisms must have arisen **(7)** _____ .

By the mid-1800s, however, **(8)** _____ was able to disprove this hypothesis once and for

all. He set up an experiment, using flasks with unique **(9)** _____ necks. These flasks

allowed **(10)** _____ , but no organisms, to come into contact with a broth containing

nutrients. If some **(11)** _____ existed, as had been suggested, it would be able to get into

the **(12)** _____ through the open neck of the flask. His experiment proved that organ-

isms arise only from other **(13)** _____ . This idea, called **(14)** _____ ,

is one of the cornerstones of biology today.

Determine if the statement is true. If it is not, rewrite the italicized part to make it true.

15. Biogenesis *explains* how life began on Earth.

_____

16. For life to begin, simple *inorganic* molecules had to be formed and then organized into complex molecules.

_____

17. Several billion years ago, Earth's atmosphere had no free *methane*.

_____

18. Primitive Earth's atmosphere may have been composed of water vapor, carbon dioxide, and

nitrogen. _____

19. In the early 1900s, Alexander Oparin proposed a widely accepted hypothesis that life began *on land.*

_____

20. *Pasteur* hypothesized that many chemical reactions occurring in the atmosphere resulted in the for-

mation of a primordial soup. _____

21. In 1953, Miller and Urey tested Oparin's hypothesis by simulating the conditions of *modern* Earth in

the laboratory. _____

22. Miller and Urey showed that organic compounds, including *nucleic acids* and sugars, could be formed

in the laboratory, just as had been predicted. _____

23. This "life-in-a-test-tube" experiment of Miller and Urey provides support for some modern hypothe-

ses of *biogenesis.* _____

24. Sidney Fox took Miller and Urey's experiment further and showed how amino acids could cluster to

form *protocells.* _____

In your textbook, read about the evolution of cells.

Answer the following questions.

25. Describe the likely characteristics of the first organisms on Earth.

_____

_____

_____

26. What is an autotroph? What factors helped them thrive on Earth?

_____

_____

_____

27. What present-day organisms may be similar to the first autotrophs? Why?

_____

_____

28. What change occurred in Earth's atmosphere after the evolution of photosynthesizing prokaryotes?
Why?

_____

_____

Chapter 15 **The Theory of Evolution**

Section 15.1 Natural Selection and the Evidence for Evolution

In your textbook, read about Charles Darwin and natural selection.

For each statement, write <u>true</u> or <u>false</u>.

_____ **1.** H.M.S. *Beagle*, upon which Charles Darwin served as naturalist, set sail on a collecting and mapping expedition in 1831.

_____ **2.** The environments that Darwin studied exhibited little biological diversity.

_____ **3.** By careful anatomical study, Darwin found that the many species of plants and animals on the Galápagos Islands were unique and bore no relation to species seen in other parts of the world.

_____ **4.** The tortoises of the Galápagos Islands are among the largest on Earth.

_____ **5.** After returning to England, Darwin studied his collections for 10 years.

_____ **6.** Darwin named the process by which evolution proceeds *artificial selection*.

You are a naturalist who traveled to the Galápagos Islands. Below are excerpts from field notes. Next to each set of notes, write a heading. Use these choices: Overproduction of Offspring, Natural Selection, Struggle for Existence, Variation.

7.

Field Notes

Female finches found on the Galápagos Islands lay enormous numbers of eggs.

_____

8.

Field Notes

These finches compete for a particular species of insect that inhabits the small holes found in tree bark.

_____

9.

Field Notes

Some finches' beaks are long, some are short. The finches with long beaks are better adapted to remove the insects from the bark.

_____

10.

Field Notes

The finches with the long beaks survive and produce greater numbers of offspring with long beaks.

_____

In your textbook, read about natural selection and adaptations.

Identify the type of structural adaptation that the statement describes. If the statement applies to both, write <u>both</u>. Use these choices: mimicry, camouflage, both.

_____ **11.** Enable(s) an organism to blend in with its surroundings

_____ **12.** Provide(s) protection for an organism by copying the appearance of another species

_____ **13.** The coloration of a flounder that allows the fish to avoid predators

_____ **14.** Involve(s) changes to the external appearance of an organism

_____ **15.** A flower that looks like a female bee

In your textbook, read about evidence for evolution.

Complete the chart by checking the kind of evidence described.

Evidence	Type of Evidence				
	Homologous Structure	**Analogous Structure**	**Vestigial Structure**	**Embryological Development**	**Genetic Comparisons**
16. A modified structure seen among different groups of descendants					
17. In the earliest stages of development, a tail and pharyngeal pouches can be seen in fish, birds, rabbits, and mammals.					
18. Exemplified by forelimbs of bats, penguins, lizards, and monkeys					
19. Eyes in a blind fish					
20. DNA and RNA comparisons may lead to evolutionary trees.					
21. Bird and butterfly wings have same function but different structures					
22. A body structure reduced in original function but may have been used in an ancestor					

In your textbook, read about population genetics and evolution.

Determine if the statement is true. If it is not, rewrite the italicized part to make it true.

1. *Adaptations* of species are determined by the genes contained in the DNA code. _____

2. When Charles *Mendel* developed the theory of natural selection in the 1800s, he did not include a genetic explanation. _____

3. Natural selection can act upon an individual's *genotype*, the external expression of genes.

_____

4. Natural selection operates on *an individual* over many generations. _____

5. The entire collection of genes among a population is its *gene frequency*. _____

6. If you know the *phenotypes* of all the organisms in a population, you can calculate the allelic frequency of the population. _____

7. A population in which frequency of alleles *changes* from generation to generation is said to be in genetic equilibrium. _____

8. A population that is in *genetic equilibrium* is not evolving. _____

9. Any factor that affects *phenotype* can change allelic frequencies, thereby disrupting the genetic equilibrium of populations. _____

10. Many *migrations* are caused by factors in the environment, such as radiation or chemicals, but others happen by chance. _____

11. Mutations are *important* in evolution because they result in genetic changes in the gene pool.

_____

12. Genetic *equilibrium* is the alteration of allelic frequencies by chance processes. _____

13. Genetic drift is more likely to occur in *large* populations. _____

14. The factor that can significantly change the genetic equilibrium of a population's gene pool is *mutation*. _____

15. The type of natural selection by which one of the extreme forms of a trait is favored is called *disruptive selection*. _____

Chapter 15 The Theory of Evolution, *continued*

In your textbook, read about the evolution of species.

Complete each statement.

16. _____ can occur only when either interbreeding or the production of fertile offspring is prevented among members of a population.

17. _____ occurs when formerly interbreeding organisms are prevented from producing fertile offspring.

18. Polyploid speciation is perhaps the fastest form of speciation because it results in immediate

    _____ .

19. The hypothesis that species originate through a slow buildup of new adaptations is known as

    _____ .

20. This hypothesis is supported by evidence from the _____ record.

21. The hypothesis of _____ states that speciation may occur rapidly.

In your textbook, read about patterns of evolution.

Answer the following questions.

22. What happened to the ancestor of the honeycreeper when it left the mainland and encountered the diverse niches of Hawaii?

    _____

    _____

23. What is adaptive radiation?

    _____

    _____

24. Adaptive radiation is one example of divergent evolution. When does divergent evolution occur?

    _____

    _____

    _____

25. When will convergent evolution occur?

    _____

    _____

Copyright © Glencoe/McGraw-Hill, a division of The McGraw-Hill Companies, Inc.

Chapter 16 Primate Evolution

Section 16.1 Primate Adaptation and Evolution

In your textbook, read about the characteristics of a primate.

Complete the chart by checking those structures or functions that are characteristic of primates.

Structure/Function	Primate
1. Round head	
2. Flattened face	
3. Small head	
4. Large relative brain size	
5. Highly developed vision	
6. Poor vision	
7. Binocular vision	
8. Color vision	
9. Color-blind	
10. Vision the dominant sense	
11. Smell the dominant sense	
12. Immobile joints	
13. Flexible shoulder joints	
14. Skeleton adapted for movement among trees	
15. Skeleton adapted for swimming	
16. Hands and feet equipped with claws	
17. Hands and feet equipped with nails	
18. Eyes face to the side	
19. Feet constructed for grasping	
20. Opposable thumbs	

In your textbook, read about primate origins.

For each statement below, write <u>true</u> or <u>false</u>.

_____ **21.** Scientists believe that primates evolved about 66,000 years ago.

_____ **22.** The earliest primate may have been a strepsirrhinelike animal called *Purgatorius*.

_____ **23.** Anthropoids are a group of small-bodied primates.

_____ **24.** Strepsirrhines include lemurs and tarsiers.

_____ **25.** Strepsirrhines can be found in the tropical forests of South America.

Chapter 16 Primate Evolution, *continued*

Identify the following pictures. Use these choices: baboon, tarsier, spider monkey. Then on the second line write the group that is represented by the picture. Use these choices: New World monkey, Old World monkey, haplorhine.

26. _____

_____

27. _____

_____

28. _____

_____

Answer the following questions.

29. What do similarities among monkeys, apes, and humans indicate about their evolution?

_____

_____

30. According to the fossil record, what were the first modern anthropoids to evolve and about when did they evolve?

_____

_____

31. What is the evolutionary history of primates based on?

_____

_____

32. What may have led to the eventual speciation of baboons and other ground-living monkeys?

_____

_____

33. What does DNA analysis of modern hominoids suggest about their evolutionary history?

_____

_____

Primate Evolution, *continued*

Reinforcement and Study Guide

Section 16.2 Human Ancestry

In your textbook, read about hominids.

Answer the following questions.

1. What is an australopithecine? _____

2. What fossil skull did Raymond Dart discover in Africa in 1924? _____

3. Why was *A. africanus* unlike any primate fossil skull that Dart had ever seen? _____

_____

4. What did the position of the foramen magnum indicate to Dart? _____

_____

Label the following skulls. Use these choices: chimpanzee, human, *A. afarensis*

5. _____ **6.** _____ **7.** _____

For each statement below, write <u>true</u> or <u>false</u>.

_____ **8.** Much of what scientists know about australopithecines comes from the "Lucy" skeleton.

_____ **9.** "Lucy" is 3.5 billion years old.

_____ **10.** "Lucy" is classified as *A. africanus*.

_____ **11.** *A. afarensis* is the earliest known hominid species.

_____ **12.** *A. afarensis* walked on all four legs and had a humanlike brain.

_____ **13.** Australopithecines are alive today and can be found in southern Africa and Asia.

_____ **14.** Australopithecines probably played a role in the evolution of modern hominids.

In your textbook, read about the emergence of modern humans.

Circle the letter of the choice that best completes the statement or answers the question.

15. The first skull of *Homo habilis* was discovered by
  a. Raymond Dart.
  b. Louis and Mary Leakey.
  c. Donald Johanson.
  d. Gert Terblance.

16. When compared to an australopithecine skull, the *Homo habilis* skull is
  a. more humanlike.
  b. less humanlike.
  c. more apelike.
  d. exactly the same.

17. Which of the following is *not* true about *Homo habilis*?
  a. They existed between 1.5 and 2 million years ago.
  b. They were the first hominids to make and use tools.
  c. They were probably scavengers of their food.
  d. They gave rise to *A. africanus.*

18. *Homo habilis* means
  a. "handy human."
  b. "tool-using human."
  c. "upright human."
  d. "talking human."

19. Of the primates below, which has the largest brain?
  a. *Homo habilis*
  b. *Homo erectus*
  c. an ape
  d. an australopithecine

20. Which of the following is *not* true about *Homo erectus*?
  a. They probably hunted.
  b. They were the first hominids to use fire.
  c. They may have given rise to hominids that resemble modern humans.
  d. They were found only in Africa.

21. *Homo sapiens* includes
  a. Neandertals.
  b. australopithecines.
  c. *A. africanus.*
  d. *A. afarensis.*

Determine whether each statement below best describes Neandertals, Cro-Magnons, or both.

_____ **22.** They lived in caves during the ice ages.

_____ **23.** They are identical to modern humans in height, skull, and teeth structure.

_____ **24.** They may have been the first hominids to develop religious views.

_____ **25.** They may have used language.

_____ **26.** They were talented toolmakers and artists.

Chapter 17 Organizing Life's Diversity

In your textbook, read about how classification began and about biological classification.

For each item in Column A, write the letter of the matching item in Column B.

	Column A	Column B
_____	**1.** Grouping objects or information based on similarities	**a.** Aristotle
_____	**2.** Naming system that gives each organism a two-word name	**b.** Linnaeus
_____	**3.** Developed the first system of classification	**c.** genus
_____	**4.** Branch of biology that groups and names organisms	**d.** classification
_____	**5.** Designed a system of classifying organisms based on their physical and structural similarities	**e.** taxonomy
_____	**6.** Consists of a group of similar species	**f.** binomial nomenclature

If the statement is true, write *true*. If it is not, rewrite the italicized part to make it true.

7. The scientific name of a species consists of a *family* name and a descriptive name.

_____

8. The scientific name of modern humans is *Homo sapiens*.

_____

9. *Latin* is the language of scientific names.

_____

10. The *scientific* names of organisms can be misleading.

_____

11. Taxonomists try to identify the *evolutionary relationships* among organisms.

_____

12. Besides comparing the structures of organisms, taxonomists also compare the organisms' geographic distribution and *chemical makeup*.

_____

13. Similarities between living species and extinct species *cannot* be used to determine their relationship to each other.

_____

14. Because the bones of some dinosaurs have large internal spaces, some scientists think dinosaurs are more closely related to *amphibians* than to reptiles.

_____

15. Classification can be useful in identifying the *characteristics* of an unknown organism.

_____

Chapter 17 **Organizing Life's Diversity,** *continued*

In your textbook, read about how living things are classified.

Examine the table showing the classification of four organisms. Then answer the questions.

Taxon	Green Frog	Mountain Lion	Domestic Dog	Human
Kingdom	Animalia	Animalia	Animalia	Animalia
Phylum	Chordata	Chordata	Chordata	Chordata
Class	Amphibia	Mammalia	Mammalia	Mammalia
Order	Anura	Carnivora	Carnivora	Primates
Family	Ranidae	Felidae	Canidae	Hominidae
Genus	*Rana*	*Felis*	*Canis*	*Homo*
Species	*Rana clamitans*	*Felis concolor*	*Canis familiaris*	*Homo sapiens*

16. Which taxon includes the most specific characteristics? _____

17. Which taxon includes the broadest characteristics? _____

18. Which taxon includes more species, an order or a family? _____

19. Which taxon includes only organisms that can successfully interbreed? _____

20. If two organisms belong to the same family, what other taxonomic groups do the organisms have in common.

_____

21. Which two organisms in the chart are most closely related? Explain.

_____

_____

_____

_____

22. To which taxa do all four organisms belong?

_____

23. Which class does not include animals that have hair or fur? _____

24. What is the order, family, and genus of a human?

_____

25. Using the information in the chart, what can you conclude about the classification taxa of an organism with the scientific name *Rana temporaria*?

_____

_____

_____

Chapter 17 **Organizing Life's Diversity,** continued

In your textbook, read about how evolutionary relationships are determined.

Explain how scientists use each item below to determine the evolutionary relationships among organisms.

1. structural similarities: _____

_____

2. breeding behavior: _____

_____

3. geographical distribution: _____

_____

4. chromosome comparisons: _____

_____

5. biochemistry: _____

_____

In your textbook, read about phylogenetic classification: models.

Use the cladogram to answer the questions.

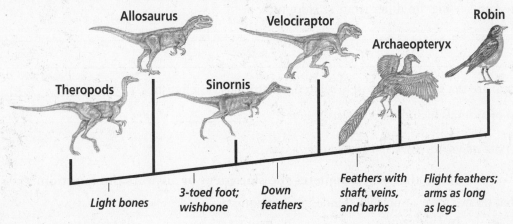

6. What five probable ancestors of the modern bird (robin) are shown on the cladogram?

_____

7. Which dinosaur is probably the most recent common ancestor of *Velociraptor* and *Archaeopteryx*?

_____

8. Which traits shown on the cladogram are shared by *Archaeopteryx* and modern birds?

_____

Chapter 17 **Organizing Life's Diversity,** *continued*

Use the fanlike phylogenetic diagram to answer the questions.

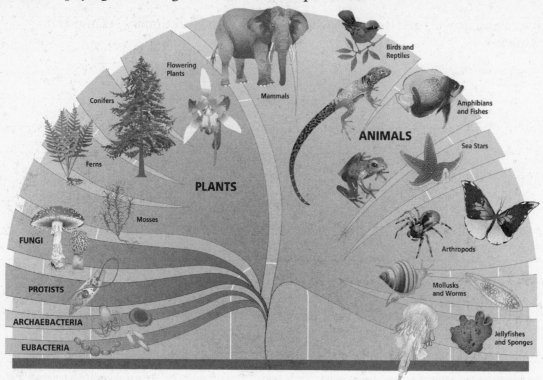

9. How does the fanlike diagram differ from a cladogram?

_____

_____

10. To which group are sea stars more closely related, arthropods or jellyfishes? _____

11. Which group of animals includes the fewest species? _____

In your textbook, read about the six kingdoms of organisms.

Circle the letter of the choice that best completes the statement or answers the question.

12. Organisms that do not have a nucleus bounded by a membrane are
   a. multicellular.   **b.** eukaryotes.   **c.** protists.   **d.** prokaryotes.

13. Fungi obtain food by
   a. photosynthesis.   **b.** chemosynthesis.
   c. endocytosis.   **d.** absorbing nutrients from organic materials.

14. Animals are
   a. autotrophs.   **b.** heterotrophs.   **c.** prokaryotes.   **d.** stationary.

BioDigest
5 **Change Through Time**

In your textbook, read about the geologic time scale.

Complete the table.

Era	Time period	Biologic event
1.	4.6 billion–600 million years ago	**2.**
3.	600 million–248 million years ago	**4.**
5.	248 million–65 million years ago	**6.**
7.	65 million years ago–present	**8.**

In your textbook, read about origin of life theories.

Complete each statement.

9. Spontaneous generation assumes that life arises spontaneously from _____ .

10. Francesco Redi and Louis Pasteur designed _____ to disprove spontaneous generation.

11. The theory of _____ states that life comes only from pre-existing life.

12. Clusters of organic molecules might have formed _____ , which may have evolved into the first true cells.

Order the evolutionary development of the following organisms from 1 to 4.

_____ **13.** chemosynthetic prokaryotes   _____ **15.** heterotrophic prokaryotes

_____ **14.** eukaryotes   _____ **16.** oxygen-producing photosynthetic prokaryotes

In your textbook, read about the evidence and mechanics of evolution.

Answer the following questions.

17. What assumption is made in the relative dating of fossils? _____

_____

18. What are homologous structures? _____

BioDigest 5 **Change Through Time,** *continued*

For each statement below, write true or false.

_____ **19.** Evolution occurs when a population's genetic equilibrium remains unchanged.

_____ **20.** Mutations, genetic drift, and migration may disrupt the genetic equilibrium of populations.

_____ **21.** Stabilizing selection favors the survival of a population's average individuals for a feature.

_____ **22.** Disruptive selection occurs when an extreme feature is naturally selected.

_____ **23.** Adaptive radiation occurs when species that once were similar to an ancestral species become increasingly distinct due to natural selection pressures.

In your textbook, read about primate evolution.

For each item in Column A, write the letter of the matching item in Column B.

	Column A		Column B
_____	**24.** Primate adaptation	**a.**	anthropoids
_____	**25.** Primate category that includes humans and apes	**b.**	Australopithecines
_____	**26.** Characteristic of New World monkeys	**c.**	genus *Homo*
_____	**27.** Appearing in fossil record about 2 million years ago along with stone tools	**d.**	opposable thumb
_____	**28.** Possible human ancestors dating from 5 to 8 million years ago	**e.**	prehensile tail

In your textbook, read about organizing life's diversity.

Look at the taxonomic classification of a bobcat shown below. Answer the questions.

29. What is the largest taxon in this classification system?

_____

30. What is the scientific name of a bobcat?

_____

Taxon	Name
Kingdom	*Animalia*
Phylum	*Chordata*
Class	*Mammalia*
Order	*Carnivora*
Family	*Felidae*
Genus	*Lynx*
Species	*rufus*

REINFORCEMENT AND STUDY GUIDE

Chapter
18 Viruses and Bacteria

Section 18.1 Viruses

In your textbook, read about the characteristics of a virus.

For each item in Column A, write the letter of the matching item in Column B.

Column A	Column B
_____ 1. Genetic material of a virus	**a.** virus
_____ 2. Where a virus attaches to a host cell	**b.** T4 phage
_____ 3. Nonliving particle that replicates inside a living cell	**c.** DNA or RNA
_____ 4. A virus's protein coat	**d.** capsid
_____ 5. Interlocks with a molecular shape in a host cell's plasma membrane	**e.** receptor site
_____ 6. Layer that surrounds the capsid of some viruses	**f.** envelope
_____ 7. A virus that infects *E. coli* bacteria	**g.** host
_____ 8. A cell in which a virus replicates	**h.** attachment protein

In your textbook, read about viral replication cycles.

Complete the table by checking the correct column for each statement.

Statement	Lytic Cycle	Lysogenic Cycle
9. Viral genes are expressed immediately after the virus infects the host cell.		
10. Many new viruses are assembled.		
11. This cycle is preceded by a virus entering a host cell.		
12. Viral DNA is integrated into the host cell's chromosome.		
13. Viruses are released from the host cell by lysis or exocytosis.		
14. Reverse transcriptase is used to make DNA from the RNA of a retrovirus.		
15. A provirus is replicated along with the host cell's chromosome.		

Use each of the terms below just once to complete the passage.

DNA   white blood cells   lysogenic
lytic   AIDS   proviruses

Many disease-causing viruses have both lytic and **(16)** _____ cycles. For example,

when HIVs infect **(17)** _____ , the viruses enter a lysogenic cycle. Their genetic

material becomes incorporated into the **(18)** _____ of the white blood cells, forming

(19) _____ . When this happens, the white blood cells still function normally, and the

person may not appear ill. Eventually, the proviruses enter a **(20)** _____ cycle, killing

the white blood cells. As a result, the person loses the ability to fight diseases and develops

(21) _____ .

In your textbook, read about viruses and cancer, plant viruses, and the origin of viruses.

If the statement is true, write *true.* **If it is not, rewrite the italicized part to make it true.**

_____ **22.** Some viruses can change normal cells to *tumor* cells.

_____ **23.** Retroviruses and the papilloma virus, which causes *hepatitis B,*
                              are examples of tumor viruses.

_____ **24.** *All* plant viruses cause diseases in plants.

_____ **25.** The first virus ever identified was the plant virus called *tobacco mosaic virus.*

_____ **26.** The patterns of color in some flowers are caused by *tumor* viruses.

_____ **27.** Tumor viruses contain genes that are found in *normal* cells.

_____ **28.** Scientists think viruses originated from *their host cells.*

Chapter 18 Viruses and Bacteria, *continued*

Section 18.2 Archaebacteria and Eubacteria

In your textbook, read about the diversity of prokaryotes and about the characteristics of bacteria.

Answer the following questions.

1. What are three types of environments in which archaebacteria are found? _____

_____

2. In what three ways do eubacteria obtain nutrients? _____

_____

3. How does a bacterium's cell wall protect it? _____

_____

4. Where is the genetic material of a bacterium found? _____

_____

5. What structure do some bacteria use to move? _____

6. What is the difference between gram-positive bacteria and gram-negative bacteria? _____

_____

_____

_____

7. What are three different shapes of bacteria? _____

8. Describe the three growth patterns of bacteria and state the prefix used to identify each growth pattern.

_____

_____

Identify the type of bacterial reproduction described. Use these choices: binary fission, conjugation.

_____ **9.** Bacterium with a new genetic makeup is produced.

_____ **10.** Circular chromosome is copied.

_____ **11.** Genetic material is transferred through a pilus.

_____ **12.** Two identical cells are produced.

_____ **13.** Sexual reproduction occurs.

Copyright © Glencoe/McGraw-Hill, a division of The McGraw-Hill Companies, Inc.

Chapter 18

Viruses and Bacteria, *continued*

Section 18.2 Archaebacteria and Eubacteria

In your textbook, read about adaptations in bacteria and the importance of bacteria.

Circle the letter of the choice that best completes the statement.

14. Scientists think the first bacteria on Earth were
    a. aerobic.   **b.** anaerobic.   **c.** fatal.   **d.** oxygen-dependent.

15. Bacteria that are obligate anaerobes release energy from food by
    a. cellular respiration.   **b.** using oxygen.
    c. using nitrogen.   **d.** fermentation.

16. As an endospore, a bacterium
    a. produces toxins   **b.** dries out.
    c. causes diseases   **d.** is protected

17. Botulism is caused by endospores of *C. botulinum* that have
    a. been killed.   **b.** produced toxins.
    c. germinated.   **d.** reproduced.

18. Nitrogen is important because all organisms need it to make
    a. proteins.   **b.** ATP.   **c.** DNA.   **d.** all of these.

19. The process by which bacteria use enzymes to convert nitrogen gas into ammonia is called
    a. nitrogenation.   **b.** atmospheric separation.
    c. nitrogen fixation.   **d.** eutrophication.

20. Bacteria return nutrients to the environment by breaking down
    a. dead organic matter.   **b.** inorganic materials.
    c. enzymes and sugar.   **d.** nitrogen in legumes.

21. Bacteria are *not* used to make
    a. vinegar.   **b.** jams.   **c.** cheese.   **d.** yogurt.

22. Bacteria are responsible for the following diseases:
    a. strep throat and tetanus.   **b.** gonorrhea and syphilis.
    c. tuberculosis and diphtheria. **d.** all of these.

23. Due to reduced death rates from bacterial diseases and improved sanitation and living conditions, the average person born in the United States today will live to be about
    a. 25 years old.   **b.** 50 years old.
    c. 75 years old.   **d.** 90 years old.

Chapter 19 Protists

In your textbook, read about what a protist is.

Determine if the statement is true. If it is not, rewrite the italicized part to make it true.

1. The kingdom *Protista* is the most diverse of all six kingdoms. _____

2. Protists can be grouped into three general types—*animal-like*, *plantlike*, and *viruslike*.

_____

3. All protists are *eukaryotes* that carry on most of their metabolic processes in membrane-bound

organelles. _____

In your textbook, read about the characteristics and diversity of protozoans.

Complete each statement.

4. The _____ protists are all unicellular heterotrophs known as protozoans.

5. Amoebas move and change their body shape by forming extensions of their plasma membranes called

_____ .

6. Amoebas use _____ to pump out excess water from their cytoplasm.

7. Most amoebas reproduce by _____ in which a parent produces one or more

identical offspring by dividing into two cells.

8. One group of protozoans are called _____ because they move by whipping

one or more flagella from side to side.

9. A paramecium moves by beating thousands of hairlike _____ .

10. When food supplies are low, paramecia may reproduce by undergoing a form of

_____ .

11. Parasitic protozoans called _____ live inside their hosts and may reproduce

by means of a spore.

12. Malaria is caused by protozoans of the genus _____ .

13. The insect that is responsible for transmitting malaria-causing protozoans to humans is the

_____ .

Reinforcement and Study Guide

Section 19.2 Algae: Plantlike Protists

In your textbook, read about what algae are and about their diversity.

For each item in Column A, write the letter of the item in Column B that completes the statement best.

Column A	Column B
_____ **1.** The characteristic common to all protists is that they are _____ .	**a.** algae
_____ **2.** Unicellular protists that are major producers of oxygen in aquatic ecosystems are _____ .	**b.** phyla
_____ **3.** Unicellular and multicellular photosynthetic protists are _____ .	**c.** pigments
_____ **4.** Most green, red, and brown algae are _____ algae.	**d.** phytoplankton
_____ **5.** Photosynthetic _____ are used to classify algae.	**e.** eukaryotic
_____ **6.** Algae are classified into six _____ .	**f.** multicellular

Identify the phylum of the alga shown below and label its parts. Use these choices: flagellum, mitochondrion, pellicle, chloroplast, nucleus, eyespot, contractile vacuole, Euglenophyta.

7. Phylum _____

8. _____

9. _____

10. _____

11. _____

12. _____

13. _____

14. _____

REINFORCEMENT AND STUDY GUIDE

Chapter
19 **Protists,** *continued*

Section 19.2 Algae: Plantlike Protists

Circle the letter of the choice that best completes the statement.

15. When diatoms that have been reproducing asexually reach about one-fourth of their original size, they

    a. die.

    c. reproduce sexually.

    b. triple in size.

    d. all of these.

16. Dinoflagellates are unicellular algae that

    a. have two flagella.

    c. have thick cellulose plates.

    b. create red tides.

    d. all of these.

17. Red algae are a kind of seaweed having pigments that absorb green, violet, and blue light waves, which allows the algae to

    a. live only in freshwater.

    c. live only in salt water.

    b. photosynthesize in limited light.

    d. both a and b.

18. The air bladders of brown algae allow the algae to

    a. breathe.

    c. float near the water's surface.

    b. reproduce.

    d. live in salt water.

19. A green alga that forms colonies is

    a. Spirogyra.

    c. Chlamydomonas.

    b. Ulva.

    d. Volvox.

In your textbook, read about alternation of generations.

Use each of the terms below just once to complete the passage.

diploid	gametophyte	alternation of generations	meiosis
haploid	sporophyte	spores	zygote

Some algae have a life cycle that has a pattern called **(20)** _____ .

These algae alternate between a(n) **(21)** _____ form that is called the

(22) _____ because it produces gametes, and a(n)

(23) _____ form called the **(24)** _____ . When the

haploid gametes fuse, they form a(n) **(25)** _____ from which the sporophyte

develops. Certain cells in the sporophyte undergo **(26)** _____ to form haploid

(27) _____ that develop into gametophytes.

Section 19.3 Slime Molds, Water Molds, and Downy Mildews

In your textbook, read about the different kinds of funguslike protists.

Use all the terms in the list below at least once to complete the concept map for funguslike protists.

cell division	three phyla	spores	cellular slime molds
plasmodial slime molds	water molds and mildew		flagellated reproductive cells

Funguslike protists

obtain energy by

decomposing organic materials

and can be divided into

1. _____

which are

2. _____

3. _____

4. _____

that reproduce by

that reproduce by

that reproduce by

5. _____

6. _____

7. _____

In your textbook, read about the origin of protists.

Answer the following question.

8. What does scientific evidence show is the relationship between protists and other groups of organisms?

_____

_____

_____

In your textbook, read about the general characteristics of fungi.

Answer the following questions.

1. What are the threadlike filaments in a multicellular fungus called? What do they form?

_____

_____

_____

2. Some hyphae are divided into individual cells by cross walls with pores in them. What are these cross walls called? What purpose do the pores serve?

_____

_____

3. What is one way that fungi are like plants and one way they are unlike plants?

_____

_____

_____

_____

In your textbook, read about adaptations in fungi.

If the statement is true, write *true*. If it is not, rewrite the italicized part to make it true.

4. Many fungi are decomposers, which *break down organic substances* into raw materials that can be used by other organisms. _____

5. Fungi use *cellular digestion* to obtain nutrients. _____

6. Hyphae release *digestive enzymes* that break down molecules in their food source.

_____

7. *Mutualistic* fungi are decomposers. _____

8. *Parasitic fungi grow spores* into host cells and absorb the cell's nutrients.

_____

In your textbook, read about reproduction in fungi.

Complete each statement.

9. Fungi reproduce _____ by fragmentation, budding, or producing spores.

10. In _____ , pieces of hyphae grow into new mycelia.

11. The process of a parent cell undergoing mitosis and producing a new individual that pinches off, matures, and separates from the parent is called _____ .

12. When environmental conditions are right, a _____ may germinate and produce a threadlike _____ that will grow into a mycelium.

13. Some hyphae grow away from the mycelium to produce a spore-containing structure called a _____ .

14. In most fungi, the structures that support _____ are the only part of the fungus that can be seen.

15. Fungi may produce spores by _____ or _____ .

16. Many adaptations of fungi for survival involve _____ .

17. _____ protect spores and keep them from from drying out until they are released.

18. A single puffball may produce a cloud containing as many as _____ spores.

19. Producing a large number of spores increases a species' chances of _____ .

20. Fungal spores can be dispersed by _____ , _____ , and _____ .

Chapter 20 **Fungi,** *continued*

In your textbook, read about zygomycotes.

Order the steps of growth and reproduction in zygomycotes from 1 to 5.

_____ **1.** Hyphae called rhizoids penetrate the food, anchor the mycelium, and absorb nutrients.

_____ **2.** An asexual spore germinates on a food source and hyphae begin to grow.

_____ **3.** Spores are released and another asexual cycle begins.

_____ **4.** Hyphae called stolons grow across the surface of the food source and form a mycelium.

_____ **5.** Special hyphae grow upward to form sporangia that are filled with asexual spores.

Use each of the terms below just once to complete the passage.

yeasts	conidia	multicellular	conidiophores	yeast cells
sac fungi	ascospores	unicellular	vaccine	ascus

Ascomycotes are also called **(6)** _____ because they produce sexual spores,

called **(7)** _____ , in a saclike structure, called a(n) **(8)** _____ .

During asexual reproduction, ascomycotes produce spores called **(9)** _____ .

These asexual spores are produced in chains or clusters at the tips of structures called

(10) _____ , which are elongated hyphae.

Morels and truffles are **(11)** _____ ascomycotes that are edible. Yeasts are

(12) _____ ascomycotes. **(13)** _____ are used to make

beer, wine, and bread. They are also used in genetic research. A(n) **(14)** _____

for the disease hepatitis B is produced from rapidly growing **(15)** _____ , which

contain spliced human genes.

Chapter 20 **Fungi,** continued

In your textbook, read about basidiomycotes.

Answer the following questions about the life of a mushroom.

16. What are basidia and where are they found?

_____

_____

_____

17. What happens when mycelia of two different mating strains meet?

_____

18. What does a mycelium with two nuclei in its cells form?

_____

19. What does a diploid cell inside a basidium produce as a result of meiosis?

_____

_____

In your textbook, read about deuteromycotes, the mutualistic relationships of mycorrhizae and lichens, and the origins of fungi.

Write the letter of the item in Column B that best completes each statement in Column A.

	Column A	Column B
_____	**20.** _____ is an example of a deuteromycote.	**a.** basidiomycotes
_____	**21.** A mycorrhiza is a mutualistic relationship between a fungus and a(n) _____ .	**b.** mycorrhizae
_____	**22.** _____ is an antibiotic produced by a deuteromycote.	**c.** alga
_____	**23.** Plants that have _____ associated with their roots grow larger.	**d.** *Penicillium*
_____	**24.** _____ make up a division of fungi that have no known sexual stage.	**e.** pioneer species
_____	**25.** A lichen is a mutualistic relationship between a fungus and a(n) _____ or cyanobacterium.	**f.** deuteromycotes
_____	**26.** Lichens are _____ in all parts of the world.	**g.** penicillin
_____	**27.** Scientists think that ascomycotes and _____ evolved from a common ancestor.	**h.** plant

BioDigest 6 Viruses, Bacteria, Protists, and Fungi

In your textbook, read about viruses.

Label the parts of a virus.

1. _____

2. _____

3. _____

Number the following steps of the lytic cycle in the order in which they occur.

_____ **4.** The viral nucleic acid causes the host cell to produce new virus particles.

_____ **5.** A virus attaches to the membrane of a host cell.

_____ **6.** The new virus particles are released from the host cell, killing the cell.

_____ **7.** The viral nucleic acid enters the host cell.

In your textbook, read about bacteria.

Determine if the statement is true. If it is not, rewrite the italicized part to make it true.

_____ **8.** A bacterium is a unicellular *eukaryote*.

_____ **9.** Bacteria may be heterotrophs, photosynthetic autotrophs, or *chemosynthetic autotrophs*.

_____ **10.** Bacteria reproduce asexually by *conjugation*.

_____ **11.** Bacteria that are *obligate aerobes* are killed by oxygen.

_____ **12.** *Archaebacteria* often live in extreme environments.

_____ **13.** Some bacteria fix *oxygen*.

BioDigest 6 **Viruses, Bacteria, Protists, and Fungi,** *continued*

In your textbook, read about protists.

Write the letter of the item in Column B that best matches the item in Column A.

Column A	Column B
_____ **14.** Can be unicellular, colonial, or multicellular	**a.** green algae
_____ **15.** Parasitic protozoans	**b.** sporozoans
_____ **16.** Causes malaria	**c.** amoeba
_____ **17.** Can be both autotrophic and heterotrophic	**d.** slime mold
_____ **18.** Funguslike protist	**e.** brown algae
_____ **19.** Uses cilia to move	**f.** *Plasmodium*
_____ **20.** Contain carotenoids	**g.** euglenas
_____ **21.** Have hard, armorlike plates	**h.** *Paramecium*
_____ **22.** Kelps	**i.** dinoflagellates
_____ **23.** Uses pseudopodia to move	**j.** diatoms
_____ **24.** Have red and blue pigments	**k.** red algae

In your textbook, read about fungi.

Answer the following questions.

25. How do fungi obtain nutrients from a food source? _____
_____

26. How do fungi play a role in recycling nutrients on Earth? _____
_____

27. What are hyphae? _____

28. In what structure do club fungi produce sexual spores? _____

29. In what structure do sac fungi produce sexual spores? _____

30. How do mycorrhizae benefit both the plants and the fungi? _____
_____
_____

31. What is a lichen? _____
_____

Chapter 21 What is a plant?

In your textbook, read about the origins and adaptations of plants.

For each answer given below, write an appropriate question.

1. **Answer:** Multicellular eukaryotes having thick cell walls made of cellulose, a protective, waterproof covering, and that can produce its own food in the form of glucose through photosynthesis

   Question: _____

2. **Answer:** The earliest known plant fossils

   Question: _____

3. **Answer:** Protective, waxy layers that cover most fruits, leaves, and stems

   Question: _____

4. **Answer:** The organ of a plant that usually traps light energy for photosynthesis, and is supported by a stem

   Question: _____

5. **Answer:** The organ that anchors a plant, and usually absorbs water and minerals

   Question: _____

In your textbook, read about alternation of generations.

Use each of the terms below just once to complete the passage.

diploid	generations	meiosis
gametes	haploid	sporophyte

The lives of all plants consist of two alternating stages, or **(6)** _____ . The

gametophyte generation of a plant is responsible for the development of **(7)** _____ .

All cells of the gametophyte, including the gametes, are **(8)** _____ . The

(9) _____ generation is responsible for the production of spores. All cells of the

sporophyte are **(10)** _____ . The spores are produced by the sporophyte plant by

(11) _____ and are, therefore, haploid.

In your textbook, read about the origin and adaptations of plants.

Circle the letter of the choice that best completes the statement.

12. The lives of _____ plants include two generations that alternate.
    a. non-seed producing
    b. seed
    c. all
    d. most

13. The generation of a plant responsible for producing gametes is the
    a. alternation of generations.
    b. gametophyte generation.
    c. sporophyte generation.
    d. seed-producing generation.

14. Gametophyte spores are _____ and sporophyte tissue cells are _____ .
    a. haploid/diploid.
    b. diploid/haploid.
    c. haploid/haploid.
    d. diploid/diploid.

15. Non-seed plants _____ that grow into gametophytes.
    a. release spores into the environment
    b. retain spores in the parent plant
    c. release seeds into the environment
    d. retain seeds in the parent plant

Answer the following questions.

16. What is the difference between vascular and nonvascular plants?

_____

_____

_____

_____

17. Some land plants produce seeds. What is their function? How do they differ from spores?

_____

_____

_____

_____

18. How do algae and land plants take in substances?

_____

_____

_____

_____

In your textbook, read about non-seed plants.

For each item in Column A, write the letter of the matching item in Column B.

Column A	Column B
_____ 1. Leaves that are found on ferns	**a.** leafy liverworts
_____ 2. Scaly structures that support male or female reproductive structures	**b.** thallose liverworts
_____ 3. Plants with a broad, flattened body that resembles a lobed leaf	**c.** fronds
_____ 4. Plants with three flattened rows of thin leaves	**d.** hornworts
_____ 5. Nonvascular plants that grow in damp, shady habitats and whose sporophytes resemble horns	**e.** cones

Complete the chart below by marking the appropriate columns for each division of plants.

Division	Vascular	Nonvascular	Non-seed Plants	Seeds in Fruits	Seeds not in Fruits
6. Hepaticophyta					
7. Anthocerophyta					
8. Bryophyta					
9. Psilophyta					
10. Lycophyta					
11. Arthrophyta					
12. Pterophyta					
13. Cycadophyta					
14. Gnetophyta					
15. Ginkgophyta					
16. Coniferophyta					
17. Anthophyta					

Reinforcement and Study Guide

Section 21.2 Survey of the Plant Kingdom

In your textbook, read about non-seed plants.

For each answer given below, write an appropriate question.

18. **Answer:** Vascular plants that have neither roots nor leaves

Question: _____

19. **Answer:** Vascular plants that have hollow, jointed stems surrounded by whorls of scalelike leaves, whose cells contain large deposits of silica

Question: _____

20. **Answer:** Plants that may be the ancestors of all plants

Question: _____

21. **Answer:** Hard-walled reproductive cells found in non-seed plants

Question: _____

22. **Answer:** Nonvascular plants that rely on osmosis and diffusion to transport water and nutrients, although some members have elongated cells that conduct water and sugars

Question: _____

In your textbook, read about seed plants.

Use each of the terms below just once to complete the passage.

anthophyta   cycadophyta   gnetophyta
coniferophyta   ginkgophyta

Seed plants are classified into five divisions. Plants from the **(23)** _____ division are

palmlike trees with scaly trunks and are often mistaken for ferns or small palm trees. There is only one

living species in the **(24)** _____ division. The members of the **(25)** _____

division are the largest, most diverse group of seed plants on Earth and are commonly known as the flow-

ering plants. Three distinct genera make up the plant division called **(26)** _____ . Species

of the **(27)** _____ division can be identified by the characteristics of their needlelike or

scaly leaves.

Chapter 22 The Diversity of Plants

Section 22.1 Nonvascular Plants

In your textbook, read about nonvascular plant divisions—bryophyta, hepaticophyta, and anthocerophyta.

Complete each statement.

1. Nonvascular plants are successful in habitats with adequate _____ .

2. The _____ generation is dominant in nonvascular plants.

3. Sperm are produced in male reproductive structures called _____ , and eggs are produced in female reproductive structures called _____ .

4. Mosses have colorless multicellular structures called _____ , which help anchor the stem to the soil.

5. Most liverworts have _____ that helps reduce evaporation of water from the plant's tissues.

6. Liverworts occur in many environments and include two groups: the _____ liverworts and the _____ liverworts.

7. One unique feature of hornworts is the presence of one to several _____ in each cell.

8. The common names for the nonvascular plants in bryophyta, hepaticophyta, and anthocerophyta are _____ , _____ , and _____ .

Circle the letter of the response that best completes the statement.

9. Nonvascular plants are not as common or as widespread as vascular plants because
   a. nonvascular plants are small.
   b. the life functions of nonvascular plants require a close association with water.
   c. nonvascular plants are limited to dry habitats.
   d. none of the above.

10. The life cycle of nonvascular plants includes an alternation of generations between a
    a. diploid sporophyte and a diploid gametophyte.
    b. haploid sporophyte and a haploid gametophyte.
    c. diploid sporophyte and a haploid gametophyte.
    d. haploid sporophyte and a diploid gametophyte.

11. Fossil and genetic evidence suggests that the first land plants were
    a. mosses.   c. liverworts.
    b. sphagnum moss.   d. hornworts.

Chapter 22 **The Diversity of Plants,** continued

In your textbook, read about the alternation of generations of non-seed vascular plants and lycophyta.

Use each of the terms below just once to complete the passage.

antheridia	leaves	sporophyte
archegonia	prothallus	strobilus
egg	reproductive cells	zygote
fertilization	sperm	

Unlike nonvascular plants, the spore-producing **(1)** _____ is the dominant generation

in vascular plants. A major advance in vascular plants was the adaptation of **(2)** _____ to

form structures that protect the developing **(3)** _____ . In some non-seed vascular

plants, spore-bearing leaves form a compact cluster called a(n) **(4)** _____. Spores are

released from this compact cluster. These spores then grow to form the gametophyte, called a(n)

(5) _____ . This structure is relatively small and lives in or on soil. The prothallus then

forms **(6)** _____ , male reproductive structures, and **(7)** _____ , female

reproductive structures. **(8)** _____ are released from an antheridium and swim through a

film of water to the **(9)** _____ in an archegonium. **(10)** _____ occurs and

a large, dominant sporophyte plant develops from the **(11)** _____ .

For each statement below, write <u>true</u> or <u>false</u>.

_____ **12.** The leafy stems of lycophytes resemble clubs, and their reproductive structures are moss shaped.

_____ **13.** The leaves of lycophytes occur as pairs, whorls, or spirals along the stem.

_____ **14.** Lycophytes are simple vascular plants with creeping leaves.

_____ **15.** The club moss is commonly called ground pine because it is evergreen and resembles a miniature pine tree.

In your textbook, read about arthrophyta and pterophyta.

Complete each statement.

16. The hollow-stemmed horsetail appears to be jointed with scalelike _____ surrounding each joint.

17. The most recognized generation of ferns is the _____ generation.

18. The _____ in most ferns is a thin, flat structure.

19. In most ferns, the main stem, called a _____ , is underground. It contains many starch-filled cells for _____ .

20. The leaves of a fern are called _____ and grow upward from the rhizome.

21. Fronds are often divided into _____ , which are attached to a central rachis.

22. Ferns were the first vascular plants to evolve leaves with branching _____ of vascular ranchis.

23. The common names for the seedless vascular plants, lycophyta, arthrophyta, and pterophyta are _____ , _____ , and _____ .

Answer the following questions on the lines provided.

24. Why are arthrophytes, or horsetails, sometimes referred to as scouring rushes?

_____

_____

_____

25. Why might you be more familiar with ferns than with club mosses and horsetails?

_____

_____

_____

_____

In your textbook, read about the seed plants—cycadophyta, gingkophyta, gnetophyta, coniferophyta, and anthophyta.

Complete each statement.

1. An _____ , or young diploid sporophyte, has food-storage organs called _____.

2. Vascular plants that produce _____ in cones are sometimes called _____ .

3. Seed plants do not require _____ for _____ .

4. The male gametophyte develops inside a structure called a(n) _____ that includes sperm cells, nutrients, and a protective outer covering.

5. The female gametophyte, which produces the egg cell, is contained within a sporophyte structure called a(n) _____ .

6. Biennials develop large storage organs and live for _____ .

7. Perennials produce flowers and seeds periodically for _____ .

8. Annual plants live for _____ .

9. _____ have one seed leaf; _____ have two seed leaves.

For each statement below, write _true_ or _false_.

_____ 10. Cycads are related to palm trees but their leaves unfurl like fern fronds.

_____ 11. There is only one species of ginkgo tree alive today.

_____ 12. Most gnetophytes today are found in the deserts or mountains of Africa, Asia, North America, and Central and South America.

_____ 13. Most conifers are evergreen plants that lose their needlelike leaves all at once and only grow in nutrient-rich soil.

_____ 14. Dropping leaves is an adaptation in deciduous plants to reduce water loss when it is less available, such as during winter.

_____ 15. Anthophytes are unique in that they are the only division of plants that produce fruits.

Chapter 23 Plant Structure and Function

Section 23.1 Plant Cells and Tissues

In your textbook, read about plant cells and tissues.

Match the definitions in Column 1 with the terms in describes from Column 2. Place the letter from Column 2 in the spaces under Column 1.

	Column 1	Column 2
_____	**1.** The most abundant kind of plant cells	**a.** apical meristem
_____	**2.** Long cells with unevenly thickened cell walls. This type of cell wall allows the cells to grow.	**b.** collenchyma
_____	**3.** Cells with walls that are very thick and rigid. At maturity, these cells often die, leaving the cell walls to provide support for the plant.	**c.** companion cell
_____	**4.** Dermal tissue that is composed of flattened parenchyma cells that cover all parts of the plant	**d.** cork cambium
_____	**5.** Openings in the cuticle of the leaf that control the exchange of gases	**e.** epidermis
_____	**6.** Cells that control the opening and closing of the stomata.	**f.** guard cells
_____	**7.** Hairlike projections that extend from the epidermis	**g.** meristem
_____	**8.** Plant tissue composed of tubular cells that transports water and dissolved minerals from the roots to the rest of the plant	**h.** parenchyma
_____	**9.** Tubular cells, with tapered ends, that have cell walls with pits through which water and dissolved minerals flow	**i.** phloem
_____	**10.** Lateral meristem that produces a tough covering for the surface of stems and roots	**j.** sclerenchyma
_____	**11.** Vascular tissue that transport sugars to all parts of the plant	**k.** sieve tube member
_____	**12.** Long, cylindrical cells through which sugars and organic compounds flow	**l.** stomata
_____	**13.** Nucleated cells that help manage the transport of sugars and other organic compounds through the sieve cells of the phloem	**m.** tracheids
_____	**14.** Areas where new cells are produced	**n.** trichomes
_____	**15.** Growth tissue found at or near the tips of roots and stems	**o.** vascular cambium
_____	**16.** Tubular cells that transport water throughout the plant. These cells are wider and shorter than tracheids.	**p.** vessel element
_____	**17.** Lateral meristem that produces new xylem and phloem cells in dicot stems and roots	**q.** xylem

Chapter 23 **Plant Structure and Function,** *continued*

In your textbook, read about roots and stems.

Label the parts of the dicot root. Use these choices:

cortex   phloem   epidermis   endodermis   xylem

Dicot Root

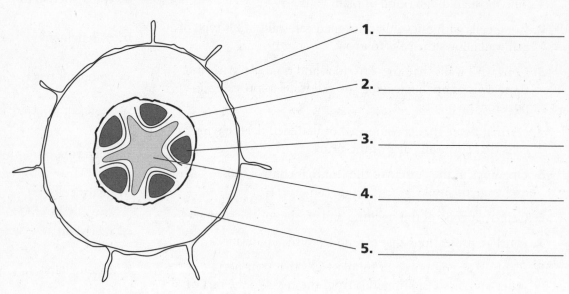

1. _____

2. _____

3. _____

4. _____

5. _____

For each statement below, write <u>true</u> or <u>false</u>.

_____ **6.** A root hair is a small extension of an epidermal, or outermost, cell layer of a dicot root.

_____ **7.** Layers of parenchyma cells make up the cortex of a dicot root and the central pith of a monocot root.

_____ **8.** Outside the endodermis is a tissue called the pericycle that develops vertical roots.

_____ **9.** Vascular cambium cells found near the center of a root produce xylem and phloem cells that increase the diameter of the root.

_____ **10.** Behind the root tip are cell-producing growth tissues called the root cap.

_____ **11.** The difference between roots and stems lies in the way they transport water.

_____ **12.** Primary growth in a stem occurs in the apical meristem.

REINFORCEMENT AND STUDY GUIDE

Chapter 23 Plant Structure and Function, *continued*

Reinforcement and Study Guide

Section 23.2 Roots, Stems, and Leaves

In your textbook, read about stems and leaves.

Circle the letter of the response that best completes the statement.

13. Many wildflowers with soft, green stems are plants that have
    a. woody stems.
    b. herbaceous stems.
    c. woody roots.
    d. all of the above.

14. The functions of a plant's stem include
    a. transporting sugar.
    b. supporting the plant.
    c. transporting water and minerals.
    d. all of the above.

15. Any portion of the plant that stores sugars is called a
    a. petiole.
    b. mesophyll.
    c. root cap.
    d. sink.

16. The movement of sugars through the phloem is called
    a. photosynthesis.
    b. transpiration.
    c. translocation.
    d. food storage.

In your textbook, read about the leaves of a plant.

Use each of the terms below just once to complete the passage.

stomata	extend	cuticle	transpiration	epidermis
veins	stem	petiole	photosynthesis	mesophyll

There are many parts to a leaf. Grass leaves grow right out of the **(17)** _____ ,

but other leaves are connected to the stem by a stalk called the **(18)** _____ .

The petiole is made of vascular tissues that **(19)** _____ up into the leaf to form

(20) _____ .

The outer surface of a leaf has a **(21)** _____ that covers the epidermis.

Inside the epidermis are two layers of photosynthetic cells that make up the

(22) _____ . Cells in the palisade layer have many chloroplasts and

carry out most of the leaf's **(23)** _____ . Leaves have a(n)

(24) _____ with a waxy cuticle and **(25)** _____ help

prevent water loss. The loss of water through stomata is called **(26)** _____ .

In your textbook, read about plant hormones and plant responses.

Complete each statement.

1. A _____ is a chemical that is produced in one part of an organism and transported to another part, where it causes a physiological change.

2. The group of plant hormones called _____ promote cell elongation. Indoleacetic acid (IAA) is an example of this group of hormones.

3. The group of growth hormones that cause plants to grow taller because, like auxins, they stimulate cell elongation, are called _____ .

4. The hormones called _____ are so named because they stimulate cell division by stimulating the production of proteins needed for mitosis.

5. The plant hormone called _____ is a simple, gaseous compound composed of carbon and hydrogen that speeds the ripening of fruits.

6. A plant's response to an external stimulus that comes from a particular direction is called a _____ .

7. A responsive movement of a plant that is not dependent on the direction of the stimulus is called a _____ .

Determine if the statement is true. If it is not, rewrite the italicized part to make it true.

8. A *large* amount of hormone is needed to make physiological changes in a plant.

_____

9. If gibberellins are applied to the tip of a dwarf plant, it will grow *taller*.

_____

10. The growth of a plant towards light is caused by an unequal distribution of *ethylene* in the plant's stem.

_____

11. If a tropism is *negative*, the plant grows toward the stimulus.

_____

12. The growth of a plant toward light is called *phototropism*.

_____

13. *Gravitropism* is the direction of plant growth in response to gravity.

_____

14. A plant's response to touch is called *cytokinin*.

_____

Chapter 24 Reproduction in Plants

Section 24.1 Life Cycles of Mosses, Ferns, and Conifers

In your textbook, read about alternation of generations and the life cycles of mosses and ferns.

Use each of the terms below just once to complete the following statements.

diploid	meiosis	sporophyte
dominant	mitosis	vegetative reproduction
egg	protonema	
gametophyte	sperm	

1. The two phases of the plant life are the _____ stage and

   the _____ stage.

2. The cells of the sporophyte are all _____ .

3. The female gamete is the _____ , and the male gamete is the _____ .

4. Some plants reproduce asexually by a process called _____ , in which a new

   plant is produced from an existing vegetative structure.

5. Mosses belong to one of the few plant divisions in which the gametophyte plant is the

   _____ generation.

6. A small, green filament of moss cells that develops into either a male or female moss gametophyte is

   known as a(n) _____ .

7. The moss diploid zygote divides by _____ to form a new sporophyte in the

   form of a stalk and capsule.

8. Spores are produced by _____ in the capsule of the moss sporophyte.

**Number each description to order the stages from spore release of the life cycle
of a fern, from 1 to 7.**

_____ 9. A spore germinates and grows into a heart-shaped gametophyte called a prothallus.

_____ 10. After fertilization, the diploid zygote grows into a sporophyte.

_____ 11. As the sporophyte grows, roots and fronds grow out from the rhizome.

_____ 12. Sperm swim through a film of water on the prothallus to reach and fertilize an egg in
the archegonium.

_____ 13. In each sporangium, spores are produced by meiosis, and the cycle begins again as the
spores are dispersed by the wind.

_____ 14. Sori, or clusters of sporangia, grow on the pinnae.

_____ 15. A sporangium bursts, releasing haploid spores.

Chapter 24 Reproduction in Plants, *continued*

Section 24.1 Life Cycles of Mosses, Ferns, and Conifers

In your textbook, read about the life cycle of conifers.

Answer the following questions.

1. What is the dominant stage in conifers?

_____

2. What does the adult conifer produce on its branches?

_____

_____

3. What is a megaspore?

_____

_____

4. What are microspores and how are they produced?

_____

_____

5. What do the microspores develop into?

_____

6. What is a micropyle?

_____

7. How does fertilization take place?

_____

_____

_____

8. After fertilization, a zygote develops inside the ovule into an embryo with several cotyledons. What happens to the ovule?

_____

_____

9. What happens to the seeds when the female cone opens and falls to the ground?

_____

_____

10. What will the seedling become?

_____

Chapter
24 **Reproduction in Plants,** *continued*

In your textbook, read about the structure of a flower.

Determine if the statement is true. If it is not, rewrite the italicized part to make it true.

1. In flowering plants, sexual reproduction takes place in the *seed*, which has several parts.

_____

2. The structure of a flower includes four kinds of organs: *sepals, petals, stamens, and ovaries.*

_____

3. *Petals* are usually colorful, leaflike structures that encircle the flower stem.

_____

4. The male reproductive structure located inside the petals of a flower is a *stamen*. Sperm-containing pollen is produced in the *anther* at the tip of the stamen.

_____

5. The female reproductive structure at the center of a flower is the *ovary*. Eggs are formed in the *pistil*, which is located in the bottom portion of the *ovary*.

_____

Label the parts of the flower. Use these choices:

sepal	petal	stigma	ovule
anther	ovary	filament	

6. _____

8. _____

10. _____

11. _____

7. _____

9. _____

12. _____

Reinforcement and Study Guide

Section 24.3 The Life Cycle of a Flowering Plant

In your textbook, read about the life cycle of a flowering plant.

For each item in Column A, write the letter of the matching item from Column B.

Column A	Column B
_____ **1.** Two nuclei in one cell at the center of the embryo sac	**a.** dormancy
_____ **2.** A process in which one sperm fertilizes the egg and the other sperm joins with the central cell	**b.** double fertilization
_____ **3.** Food-storing tissue that develops from the triploid central cell and supports the development of the embryo	**c.** endosperm
_____ **4.** A period of inactivity in which seeds of some plant species remain until conditions are favorable for growth and development	**d.** germination
_____ **5.** The beginning of the development of the embryo into a new plant	**e.** hypocotyl
_____ **6.** This embryonic root is the first part of the embryo to appear from the seed	**f.** polar nuclei
_____ **7.** The portion of the stem near the seed	**g.** radicle

Answer the following questions.

8. How do anthophytes attract animal pollinators?

_____

_____

_____

9. How do seeds form after fertilization takes place?

_____

_____

_____

_____

10. Name three ways seeds are dispersed.

_____

_____

_____

_____

BioDigest 7 Plants

In your textbook, read about plants.

Study the following diagram of alternation of generations in plants. Then answer the questions.

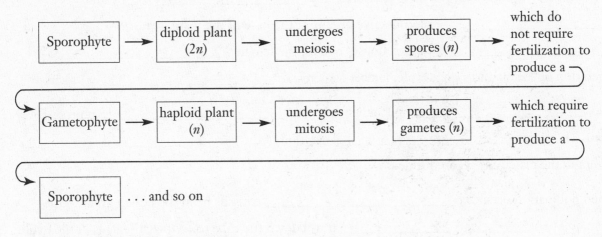

1. Is the sporophyte generation a haploid or diploid generation? _____

2. Is the gametophyte generation a haploid or diploid generation? _____

3. Is a spore haploid or diploid? _____

4. Is a gamete haploid or diploid? _____

5. Does a spore produce a gametophyte without fertilization? _____

6. Does a gamete produce a sporophyte without fertilization? _____

7. Which generation—sporophyte or gametophyte—produces a generation that is diploid?

_____

Seed Plants

Explain how these adaptations enable conifers to survive in cold or dry climates.

8. *Needles* _____

_____

9. *Stems* _____

_____

10. *Flexible leaves and branches* _____

_____

_____

Flowering Plants

Fill in the following blanks to explain the function of a flower.

11. A flower has two major reproductive structures. The _____ is the

_____ reproductive organ. At the base of the pistil is the

_____ , which houses ovules, the female _____

generation of the plant. In each ovule, female gametes, or _____ , form.

12. The _____ and anther form the _____ reproductive

organ. The male _____ generation of the plant is _____ .

Within it, the male gametes are formed.

13. When pollination occurs, a _____ extends from the pollen grain to the

ovary, and two _____ travel down the tube to fertilize the eggs in the ovule.

14. Some flowers are colorful and showy. Others are small and inconspicuous. Explain how these two
flower types are adapted to different pollinators.

_____

_____

_____

15. Discuss three ways that seeds may be spread through the environment.

   a. _____

   b. _____

   c. _____

Chapter
25 What is an animal?

In your textbook, read about the characteristics of animals.

Answer the following questions.

1. You have just discovered a new organism that you think is an animal. In order to be classified as an animal, what characteristics must it have?

   _____

   _____

2. What is one important factor that influences how an animal obtains its food?

   _____

3. How might an animal be free-moving at one stage in its life and sessile at another? Give an example.

   _____

   _____

4. How do sessile, aquatic animals get their food?

   _____

   _____

If the animal described below is a sessile organism, write yes. If it is not, write no.

_____ **5.** Barnacles attached to a ship's hull

_____ **6.** A spider lying in wait in the center of its web

_____ **7.** Coral larvae drifting in a tropical ocean

_____ **8.** Sponges growing on the outside of a crab's shell

Complete each statement.

9. Digestion in a sponge takes place in _____ _____ , while digestion in a more complex animal like a tiger takes place in a(n) _____ _____ .

10. Some of the food you had for breakfast has been stored as _____ or _____ , ready to be used if it's a long time until your next meal.

In your textbook, read about the development of animals.

If the statement is true, write <u>true</u>**. If it is not, rewrite the italicized part to make it true.**

11. Most animals develop from a single, fertilized egg called a *blastula*. _____

12. A zygote divides by a process known as *cleavage*. _____

13. The blastula is a *solid ball of cells*. _____

_____

Label the parts of the gastrula shown here. Use these choices:

ectoderm   endoderm   mesoderm   opening in gastrula

14. _____

15. _____

16. _____

17. _____

Complete the chart by checking the correct column for each description.

Description	Endoderm	Ectoderm	Mesoderm
18. Gives rise to digestive tract			
19. Continues to grow and divide			
20. Lines the inner surface of gastrula			
21. Gives rise to muscles			
22. Develops into skin and nervous tissue			
23. Forms from cells that break off endoderm			

In your textbook, read about kinds of symmetry in animals.

Circle the letter of the choice that best completes the statement or answers the question.

1. Different kinds of symmetry make it possible for animals to
- **a.** grow very large.
- **b.** survive when cut into pieces.
- **c.** move and find food in different ways.
- **d.** live a long time.

2. The irregularly shaped body of a sponge is an example of
- **a.** asymmetry.
- **b.** gastrulation.
- **c.** symmetry.
- **d.** balance.

3. A sponge's body has how many layers of cells?
- **a.** one
- **b.** two
- **c.** three
- **d.** four

4. The embryonic development of a sponge does *not* include which of the following?
- **a.** formation of endoderm
- **b.** formation of mesoderm
- **c.** a gastrula stage
- **d.** **a, b,** and **c**

5. If you divided a radially symmetrical animal along any plane through its central axis, you would end up with
- **a.** roughly equal halves.
- **b.** front and back halves.
- **c.** top and bottom halves.
- **d.** three pieces.

6. Which of the following animals is *not* radially symmetrical?
- **a.** a hydra
- **b.** a sea urchin
- **c.** a spider
- **d.** a sea star

7. An organism with bilateral symmetry can be divided lengthwise into right and left halves that are
- **a.** asymmetrical.
- **b.** similar.
- **c.** made up of two cell layers.
- **d.** flattened.

Identify each of the following body parts as being either dorsal or ventral on the animal's body.

_____ **8.** the navel of a killer whale

_____ **9.** the sail fin on an iguana

_____ **10.** the back of your neck

_____ **11.** the mouth of a shark

_____ **12.** the pouch of a kangaroo

In your textbook, read about bilateral symmetry and body plans.

Answer the following questions.

13. In what ways was the development of a body cavity, or coelom, an advantage for bilaterally symmetrical animals?

_____

_____

14. Describe an acoelomate animal's body plan.

_____

_____

15. How do nutrients get to the cells in a flatworm's solid, acoelomate body?

_____

Use each of the terms below just once to complete the passage.

coelom	completely	double	internal organs
mesoderm	partly	pseudocoelom	

A roundworm has a **(16)** _____ , a fluid-filled body cavity that is **(17)** _____

lined with **(18)** _____ . Coelomate animals have a **(19)** _____ , a body

cavity that is **(20)** _____ surrounded by mesoderm and in which complex

(21) _____ are suspended by **(22)** _____ layers of mesoderm tissue.

In your textbook, read about animal protection and support.

For each statement below, write <u>true</u> or <u>false</u>.

_____ **23.** During the course of evolution, animal body plans have decreased in complexity.

_____ **24.** An exoskeleton provides protection and support on the outside of an animal's body, as well as a place for muscle attachment.

_____ **25.** An endoskeleton is a support framework housed within the body, a protective enclosure for internal organs, and a brace for muscles to pull against.

_____ **26.** An invertebrate is an animal with a backbone.

In your textbook, read about sponges.

Answer the following questions.

1. How does the name *Porifera* relate to the structure of a sponge?

_____

_____

2. How do sponges obtain food from their environment?

_____

_____

3. Describe a sponge's body plan.

_____

_____

Complete the table by writing a cell type or structure in sponges that fits each description.

Type of Cell or Structure	Description
4.	Aid in reproduction and nutrient transport Help produce spicules
5.	Form the outside surface of body Contract to close pores
6.	Line interior of sponge's body Use flagella to draw water through pores
7.	Found in jellylike substance between layers Make up sponge's support system

Use each of the terms below just once to complete the passage.

external buds   eggs   hermaphroditic   internal fertilization

larvae   sexual   sperm

Sponges sometimes reproduce asexually by forming **(8)** _____ . Being

(9) _____ , a sponge can also produce both **(10)** _____ and sperm. During

(11) _____ reproduction, **(12)** _____ from one sponge fertilize the eggs

of another. Fertilization can be external, but **(13)** _____ is more common.

Free-swimming **(14)** _____ settle and develop into sessile adults.

Reinforcement and Study Guide

Section 26.2 Cnidarians

In your textbook, read about cnidarians.

Identify each of the following descriptions as either the <u>polyp</u> or <u>medusa</u> form of a cnidarian.

_____ **1.** Reef-building corals on the Great Barrier Reef

_____ **2.** *Aurelia*, the moon jellyfish

_____ **3.** Deep sea anemones with meter-long tentacles

_____ **4.** The asexual phase in a jellyfish's life cycle

Answer the following questions.

5. Nematocysts are characteristic of cnidarians. How does a nematocyst work?

_____

_____

6. Compare and contrast how food is digested in a sponge and in a cnidarian.

_____

_____

7. How does a nerve net function?

_____

_____

Order the following steps in the life cycle of a jellyfish from A to F, beginning with the release of eggs and sperm.

_____ **8.** A polyp grows and buds repeatedly.

_____ **9.** External fertilization takes place in the sea.

_____ **10.** A zygote develops into a blastula, which develops into a larva.

_____ **11.** Male and female medusae release sperm and eggs, respectively.

_____ **12.** A cilia-covered larva settles onto a surface.

_____ **13.** A tiny medusa breaks free from its sessile parent and drifts away.

In your textbook, read about flatworms.

For each statement below, write <u>true</u> or <u>false</u>.

_____ **1.** Flatworms are bilaterally symmetrical and have a clearly defined head.

_____ **2.** Adult planarians can focus well enough with their eyespots to form images of objects in their environment.

_____ **3.** Flame cells play an important role in maintaining water balance in planaria.

_____ **4.** A planarian uses its pharynx to locate food.

_____ **5.** Planarians reproduce sexually by producing encapsulated zygotes that hatch into free-swimming larvae.

In part C of the illustration below, draw in what you think will happen to the two halves of the cut planarian. Then, answer the question.

6.

A   B   C

7. How is regeneration adaptive for survival in planarians?

_____

_____

_____

_____

Complete the table by checking the correct column for each description.

Description	Planarian	Tapeworm	Fluke
8. Lives parasitically within a host			
9. Body made up of proglottids			
10. Body is thin and solid			
11. Free-living in aquatic environments			
12. Attaches to host's intestine with scolex			
13. Extends a pharnyx to suck up food			
14. May live in host's blood vessels			

In your textbook, read about roundworms.

Answer the following questions.

1. What impact do parasitic roundworms have on other organisms?

_____

_____

2. List three ways in which roundworms differ from flatworms.

_____

_____

3. What accounts for the characteristic wriggling movement of roundworms?

_____

_____

4. What are four of the most common parasitic roundworms that infect humans?

_____

_____

5. Can roundworms cause plant diseases? Explain.

_____

_____

6. What parts of plants are most commonly susceptible to parasitic roundworms?

_____

_____

Below are two medical reports. After reading each report, give a preliminary diagnosis of what you think might be causing the problem.

7.

MEDICAL REPORT

Patient is an active 5-year-old girl. Complains about a constant itching around the anal area, especially at night.

Preliminary Diagnosis:

_____

8.

MEDICAL REPORT

Patient is a 29-year-old female Peace Corps volunteer. Lived with remote tribe whose primary food is pigs. Complains of muscle pain.

Preliminary Diagnosis:

_____

Chapter 27 Mollusks and Segmented Worms

In your textbook, read about what a mollusk is.

The phylum *Mollusca* is a very diverse group of animals. Complete the table by checking the correct column for each characteristic.

Characteristic	Exhibited In:	
	All Mollusks	**Some Mollusks**
1. Possess a hard, external shell		
2. Bilaterally symmetrical		
3. Have a mantle		
4. Live on land		
5. Digestive tract has two openings.		
6. Inhabit aquatic environments		
7. Share similar developmental patterns		
8. Are slow-moving		
9. Have a coelom		

In your textbook, read about diversity of mollusks.

Identify each mollusk shown below. Write the name of the class to which it belongs and briefly describe where it lives.

10.

_____

_____

_____

_____

11.

_____

_____

_____

_____

12.

_____

_____

_____

_____

In your textbook, read about a mollusk's body systems and the diversity of mollusks.

Complete each statement.

13. Gastropods have either a(n) _____ shell or _____ shell.

14. Most mollusks have a(n) _____ circulatory system in which blood flows through _____ into open _____ around tissues and organs.

15. Most mollusks use _____ for respiration, while a garden slug uses a primitive _____ for gas exchange.

16. _____ are involved in removing wastes from a mollusk's body.

17. Fertilization in most aquatic mollusks takes place _____ .

For each statement below, write _true_ or _false_.

_____ **18.** All shelled gastropods are predators.

_____ **19.** When a snail is disturbed, it pulls its body inside its shell for protection.

_____ **20.** Without a shell, terrestrial slugs and sea slugs (nudibranchs) have no protection against predators.

_____ **21.** The two shells of bivalve mollusks are held together by the mantle.

Determine whether each of the statements below best describes bivalves, gastropods, or both.

_____ **22.** Nearly all feed by filtering particles from the water around them.

_____ **23.** Most have a large muscular foot.

_____ **24.** They use a radula for feeding.

_____ **25.** Water flows through their bodies via well-developed incurrent and excurrent siphons.

In your textbook, read about cephalopods.

Answer the following questions.

26. Describe the "head-foot" region of a cephalopod.

_____
_____
_____
_____

27. What would you expect to find on the interior surfaces of a squid's many arms?

_____
_____

28. How does the intelligence of an octopus compare to that of a clam?

_____
_____
_____

Using what you know about the three major classes of mollusks, complete the chart below by checking the correct column(s) for each characteristic.

Characteristics	Type of Mollusk		
	Gastropods	Bivalves	Cephalopods
29. Intelligent, with a well-developed nervous system			
30. Have no distinct head			
31. Have an open circulatory system			
32. External shells present in some species			
33. All species are carnivorous predators.			
34. Use a radula in feeding			
35. All use gills for both respiration and food collection.			
36. Bite prey with a beak			

Chapter 27 Mollusks and
Segmented Worms, continued

In your textbook, read about segmented worms, including the Inside Story about earthworms.

Use each of the terms below just once to complete the passage:

Annelida	bristleworms	earthworms	muscles
parapodia	segments	setae	

Members of the phylum **(1)** _____ all have bodies made up of multiple

(2) _____ . Each segment has its own **(3)** _____ that function to lengthen

and shorten the worm's body. When present, bristlelike **(4)** _____ act as anchors while the

worm is moving along. In **(5)** _____ , each segment has a pair of **(6)** _____ .

The most familiar annelids are probably **(7)** _____ .

If the statement is true, write *true*. If it is not, rewrite the italicized part to make it true.

8. Earthworms have a *mouth with tiny teeth* in which food particles are ground up before entering

the digestive tract. _____

9. Blood is pumped throughout an earthworm's closed circulatory system by *an elongated,*

four-chambered heart. _____

10. Some body segments in annelids are *specialized for reproduction.* _____

Below are the field notes of a biologist studying several newly collected annelid worms. Write the type of annelid—<u>earthworm</u>, <u>bristleworm</u>, or <u>leech</u>—being described.

11. Collected in rain forest of Papua, New Guinea; very active; flattened, with 32 body segments; has suckers on the ends of its body; no setae

Type of annelid: _____

12. Found crawling over corals on a reef; contains only eggs; no male reproductive organs; well-developed parapodia

Type of annelid: _____

13. Uncovered in top layer of moist soil; body has minute setae on ventral surface of each segment; hermaphroditic

Type of annelid: _____

In your textbook, read about what an arthropod is and exoskeletons.

Answer the following questions.

1. What is the most distinguishing arthropod characteristic?

_____

2. Explain the advantage of having appendages with joints.

_____

_____

_____

3. List three functions of an arthropod exoskeleton.

_____

_____

_____

In your textbook, read about molting, segmentation, and gas exchange.

Complete each statement.

4. Prior to molting, a new exoskeleton forms _____ the old one.

5. Many arthropods have three distinct body sections: a(n) _____ ,

a(n) _____ , and a(n) _____ .

6. In arthropods that have a _____ , the head and thorax are fused.

Complete the table by checking the correct column to indicate the respiratory structure you would expect to find in each example.

Example	Type of Respiratory Structure		
	Book Lungs	Gills	Tracheal Tubes
7. freshwater crayfish			
8. tarantula			
9. hissing cockroach			
10. swallowtail butterfly			

Chapter 28 **Arthropods,** *continued*

Section 28.1 Characteristics of Arthropods

In your textbook, read about arthropods' senses, body systems, and reproduction.

Identify the following as characteristics of either simple or compound eyes.

_____ **11.** have multiple lenses

_____ **12.** well-adapted for detecting slight movements

_____ **13.** have a single, focusing-type lens

_____ **14.** produce an image made up of thousands of parts

If the statement is true, write *true*. If it is not, rewrite the italicized part to make it true.

15. Animals produce pheromones, or *low frequency sounds*, that affect the behavior of others.

_____

16. In many arthropods, large, fused ganglia act as nervous system control centers *for the entire body*.

_____

17. Arthropods have an open circulatory system, in which blood leaves vessels and *comes in direct contact with body tissues*. _____

18. *Respiration* occurs in arthropods via the Malpighian tubules. _____

19. During parthenogenesis, *fertilized eggs* develop into offspring. _____

Section 28.2 Diversity of Arthropods

In your textbook, read about arthropod origins.

Answer the following questions.

20. What are the major reasons for the widespread success of arthropods?

_____

_____

21. From what animal group did arthropods probably evolve?

_____

22. List three adaptations that have evolved in arthropods.

_____

_____

_____

In your textbook, read about arachnids.

Circle the letter of the response that best completes the statement.

1. An animal that is *not* a member of the class Arachnida is
   a. a spider.   **b.** a deer tick.   **c.** a walking stick.   **d.** a dust mite.

2. In spiders, chelicerae are highly modified appendages that are adapted for
   a. holding food and injecting poison.   **b.** spinning silk and weaving webs.
   c. chewing food.   **d.** mating and reproduction.

3. The appendages of a spider that function as sense organs are
   a. its chelicerae.   **b.** its pedipalps.   **c.** its legs.   **d.** its spinnerets.

4. After catching their prey and injecting it with poison, spiders
   a. eat the prey whole.
   b. lay their eggs in the prey.
   c. chew the prey into small pieces.
   d. suck up the prey's contents, which have been liquified with enzymes.

5. In ticks and mites, the head, thorax, and abdomen
   a. are absent.   **b.** are well-defined.
   c. are fused into one section.   **d.** are all the same size.

6. The fact that horseshoe crabs have remained relatively unchanged for 500 million years indicates that
   a. natural selection has not taken place.   **b.** they must reproduce by parthenogenesis.
   c. they have very little genetic diversity.   **d.** their environment has changed very little.

In your textbook, read about crustaceans, centipedes, and millipedes.

Determine if each statement is <u>true</u> or <u>false</u>.

_____ **7.** Having compound eyes on movable stalks is an advantage for aquatic crustaceans whose potential predators could attack from almost any direction.

_____ **8.** The legs of most crustaceans are unspecialized and used only for walking.

_____ **9.** You might be more likely to see pill bugs moving around out in the open on a rainy day than on a sunny one.

_____ **10.** Both centipedes and millipedes have book lungs for gas exchange.

In your textbook, read about insects.

Using the choices below, label the diagram of a honeybee.

antennae   compound eye   legs   mandibles   spiracles   wings

11. _____

12. _____

13. _____

14. _____

15. _____

16. _____

Complete the table by checking the correct column for each statement.

Description	Type of Metamorphosis	
	Complete	Incomplete
17. Insect begins life as a fertilized egg.		
18. Larva hatches from an egg.		
19. Nymph repeatedly molts and increases in size.		
20. Nymph hatches from an egg.		
21. Pupa undergoes changes while encased in cocoon.		
22. Adults and young usually eat the same food.		
23. Adults are the only sexually mature form.		

Chapter 29 **Echinoderms and Invertebrate Chordates**

In your textbook, read about echinoderms' internal skeleton, radial symmetry, and the water vascular system.

Answer the following questions.

1. Describe the "spiny skin" that is a characteristic of echinoderms.

_____

_____

2. In what way is being radially symmetrical an advantage for adult echinoderms?

_____

_____

_____

_____

For each item in Column A, write the letter of the matching item in Column B.

Column A	Column B
_____ **3.** Has a flattened, immovable endoskeleton made up of fused plates	**a.** brittle star
_____ **4.** Has thin, flexible rays made up of small, overlapping, calcified plates	**b.** sea star
_____ **5.** Has a flexible endoskeleton divided into rather long, tapering rays	**c.** sand dollar
_____ **6.** Has tiny, calcified plates embedded in fleshy skin	**d.** sea lily
_____ **7.** Has feathery, branching rays made up of tiny, calcified plates	**e.** sea cucumber

Complete the following sentences.

8. Tube feet are part of an echinoderm's _____ _____ _____ ,

which is involved not only in locomotion, but also in _____ _____ ,

_____ , and food collecting.

9. In a sea star, water enters and exits the water vascular system through a structure called the

_____ , a sievelike, disc-shaped opening on the _____ side of the body.

In your textbook, read about sea star structure, echinoderm larvae, nutrition, nervous systems, and origins.

Label this drawing of a sea star and of a cross section of one of its rays. Use these choices:

ampulla	eyespot	madreporite	pedicellariae	tube foot

10. _____

11. _____

12. _____

13. _____

14. _____

Identify each of the following as describing either larva or an adult echinoderm.

_____ **15.** free-swimming

_____ **16.** bilaterally symmetrical

_____ **17.** radially symmetrical

_____ **18.** moves with tube feet

Determine if each of the following statements is <u>true</u> or <u>false</u>.

_____ **19.** If a sea urchin population underwent a population explosion, you might expect to see a rapid decline in the amount of algal life in the area.

_____ **20.** Sea stars and brittle stars both eat suspended organic particles.

_____ **21.** Most echinoderms have highly developed sense organs.

_____ **22.** The fact that echinoderms have deuterostome development is strong evidence that they are most closely related to chordates.

In your textbook, read about the diversity of echinoderms.

Answer the following questions.

23. List the six classes of living echinoderms and the types of animals in each class.

_____

_____

_____

24. How is the ability to regenerate lost body parts adaptive for most echinoderms?

_____

_____

_____

Complete the table by checking the column(s) that best fit(s) each description.

Description	Asteroidea	Ophiuroidea	Echinoidea	Holothuroidea	Crinoidea
25. Have multiple rays					
26. May rupture and release internal organs when threatened					
27. Some members of the class are sessile					
28. Burrow into rock or sand					
29. Use mucus-coated tentacles for feeding					
30. Some members of the class can actively swim from place to place					
31. Use rays, not tube feet, for locomotion					
32. The most inflexible type of echinoderm					
33. Use long, feathery arms to trap food particles drifting past					
34. Eat bivalves and other small animals					

Chapter 29 Echinoderms and Invertebrate Chordates, *continued*

In your textbook, read about invertebrate chordates.

Complete the following sentences.

1. At some time in their life, all chordates possess a _____ , a dorsal hollow _____ _____ , _____ _____ , and a postanal tail.

2. During your early development, your notochord became your _____ , and your pharyngeal pouches disappeared.

3. The _____ _____ is derived from the _____ portion of the dorsal nerve cord, whereas the _____ is derived from the anterior portion.

4. At some time during their lives, all chordates have a muscular _____ .

In your textbook, read about tunicates and lancelets.

Trace the path of water through a tunicate, starting with water entering the animal's body, by numbering the following statements from 1 to 5.

_____ **5.** Water leaves the pharynx region.

_____ **6.** Water passes through the gill slits, which filter food out of the water.

_____ **7.** Water is drawn into the body through the incurrent siphon.

_____ **8.** Water passes out of the body via the excurrent siphon.

_____ **9.** Water enters the pharynx, where the gill slits are located.

Complete the table by checking the correct column(s) for each description.

Description	Tunicates	Lancelets
10. Only larval forms have a tail		
11. Are filter feeders		
12. Retain all chordate traits throughout life		
13. Blood flow is continually reversed in the adult body		
14. Capable of actively swimming as adults		

BioDigest

8 Invertebrates

In your textbook, read about invertebrates.

Study the definitions on the next page and write the terms in the appropriate spaces in the crossword puzzle below. All terms are important in the BioDigest.

Copyright © Glencoe/McGraw-Hill, a division of The McGraw-Hill Companies, Inc.

BioDigest 8 **Invertebrates,** *continued*

ACROSS

1. In a animal with _____ symmetry, the right and left sides are mirrors of each other.

7. An echinoderm has an inner skeleton; its _____ covering is called the epidermis.

9. Bivalves, gastropods, and cephalopods are _____ .

10. Some echinoderms have long _____ that are used for locomotion.

12. The _____ have radial symmetry and a water vascular system.

16. Because arthropods have _____ , fossil arthropods are frequently found.

19. _____ such as planaria have no body cavity.

21. In earthworms, internal _____ are suspended from the mesoderm.

22. In some invertebrates, an exoskeleton offers _____ and support for internal tissues.

25. A(n) _____ belongs to the phylum Cnidaria.

26. Sponges have a(n) _____ body shape.

27. In earthworms and other segmented worms, each _____ has its own muscles.

28. Some segments in chordates have been modified into stacked layers called (2 words) _____ .

29. In arthropods like grasshoppers, a set of jointed appendages called antennae are adapted to give the insect acute _____ .

DOWN

1. The mouthparts of an arthropod may be adapted for such things as chewing, lapping, or _____ .

2. Echinoderm _____ have bilateral symmetry, while adults have radial symmetry.

3. _____ go through metamorphosis during their life cycles.

4. A dorsal nerve _____ is a hollow, fluid-filled canal lying above the notochord.

5. Setae, or small bristles, help earthworms with _____ .

6. Mesoderm differentiates into _____ , circulatory vessels, and reproductive organs.

8. Echinoderms have a supporting _____ , which is inside of the body instead of outside.

11. The _____ of a mollusk such as a clam is secreted by the mantle.

13. A(n) _____ functions as a watery skeleton against which muscles can work.

14. Many _____ are parasitic, such as plant-parasitic nematodes.

15. Arthropods are characterized by having a wide variety of _____ for feeding.

17. The _____ of a cnidarian is found in a highly specialized stinging cell.

18. _____ are made up of two cell layers and have only one body opening.

20. Special _____ feet enable sea stars to move from place to place.

22. Water enters a sponge through _____ .

23. A jellyfish, a(n) _____ , and an anemone are types of cnidarians.

24. Bivalves acquire food by filtering water through their _____ .

26. A radula is a tonguelike organ used by snails to scrape _____ from surfaces.

Copyright © Glencoe/McGraw-Hill, a division of The McGraw-Hill Companies, Inc.

In your textbook, read about what is a fish.

Complete each statement.

1. All vertebrates are in the phylum _____ , and fish, amphibians, reptiles, birds, and mammals are in the subphylum _____ .

2. Vertebrates are bilaterally symmetrical, coelomate animals with endoskeletons, closed _____ systems, complex _____ , and efficient _____ systems.

3. If you compared the number of fish species to the number of all other vertebrate species combined, there would be more species of _____ .

Complete the table below to compare the three different kinds of fishes.

Class	Kind of Fish	Jaws?	Skeleton	Fertilization
4. _____	Lamprey and Hagfish	**5.** _____	**6.** _____	**7.** _____
Chondrichthyes	**8.** _____	**9.** _____	Cartilage	**10.** _____
11. _____	**12.** _____	Yes	**13.** _____	External and Internal

Answer the following questions.

14. How does a fish breathe through its gills?

_____

_____

_____

_____

15. What two adaptations of cartilaginous and bony fishes help them to locate food?

_____

_____

Chapter 30 **Fishes and Amphibians,** *continued*

In your textbook, read about bony fishes.

If the statement is true, write <u>true</u>. If it is not, rewrite the italicized part to make it true.

16. While spawning, a female bony fish may lay *millions of eggs* to be fertilized externally by the male, but only a few will survive. _____

17. The development of *bone* was an important event in the evolution of vertebrates because it eventually allowed them to move onto land and support their body weight. _____

18. The backbone, comprised of separate *gills*, was a major evolutionary event for fishes because it provided support as well as flexibility, which helped to propel them through the water. _____

19. The *swim bladder*, an organ found in bony fishes, allows fishes to control their depth in water.

_____

In your textbook, read about origins of fishes.

Examine the phylogenetic tree below. Then answer the question that follows.

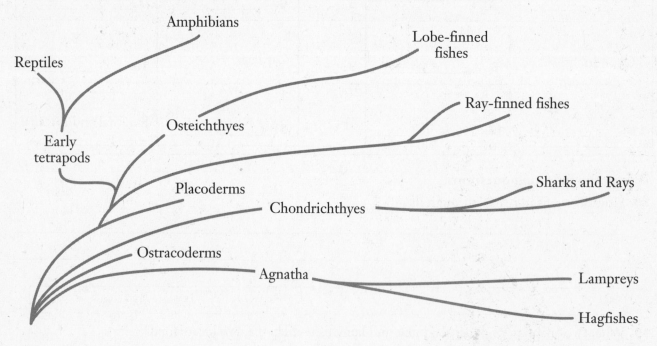

20. Which group of fishes do you think are most closely related to ancestral amphibians? Why?

_____

_____

In your textbook, read about what is an amphibian.

Answer the following questions.

1. What three orders make up the class Amphibia?

_____

_____

2. Why do amphibian eggs need to be laid in water?

_____

_____

3. Where does an amphibian heart pump oxygen-rich blood, and where does it pump oxygen-poor blood?

_____

_____

_____

_____

For each item in Column A, write the letter of the matching item in Column B.

Column A	Column B
_____ **4.** Adult frogs and toads have legs, lungs, and a _____ heart.	**a.** two-chambered
_____ **5.** Tadpoles have gills, fins, and a _____ heart.	**b.** three-chambered
_____ **6.** Amphibians are _____ , animals whose body temperature changes with the temperature of their surroundings.	**c.** skin
_____ **7.** Fertilized amphibian eggs hatch into _____ during the aquatic phase of their life.	**d.** tadpoles
_____ **8.** Some salamanders have no lungs and breathe through their _____ .	**e.** ectotherms

In your textbook, read about the characteristics and diversity of amphibians.

Circle the letter of the response that best completes the statement.

9. Early amphibians needed large amounts of food and oxygen to
   a. walk on land.
   b. breathe on land.
   c. become dormant in cold weather.
   d. all of these.

10. In many amphibians, the most important organ for gas exchange is the
   a. blood.
   b. skin.
   c. lungs.
   d. circulatory system.

11. Many frogs and toads use _____ as a defense against predators.
   a. toxins
   b. electricity
   c. sharp claws
   d. all of these

12. Frogs and toads have sound-producing bands of tissues in their throat called
   a. tongues.
   b. vocal cords.
   c. vocal tissue.
   d. none of these.

13. Salamanders are unlike frogs and toads because they have
   a. long, slender bodies.
   b. tails.
   c. necks.
   d. all of these.

14. Caecilians are amphibians that have no
   a. eyes.
   b. skin.
   c. limbs.
   d. heart.

In your textbook, read about the origins of amphibians.

For each statement below, write <u>true</u> or <u>false</u>.

_____ 15. Ancestral amphibians appeared on Earth about the same time as ancestral fishes.

_____ 16. Amphibians probably evolved from tetrapods during the Paleozoic Era.

_____ 17. Because the climate was hot and dry when amphibians first appeared on Earth, they had to stay near water.

_____ 18. Like present-day salamanders, early amphibians probably had legs set at right angles to the body.

_____ 19. Because amphibians were a transitional group, they never were the dominant vertebrates on land.

In your textbook, read about what is a reptile and the amniotic egg.

Complete the following table about reptilian adaptations and their advantages by writing in the missing information in each case.

Adaptation	Advantage
1.	In crocodilians, oxygenated and deoxygenated blood kept separate; higher level of energy production
2. Thick, scaly skin	
3.	Water not necessary for fertilization
4. Legs positioned for walking and running on land	
5.	Water not necessary for reproduction; young not overly vulnerable to aquatic predators; prevents injury or dehydration of embryo

Label the diagram below, using these choices:

albumen   allantois   amnion   chorion   embryo   shell   yolk sac

6. _____

7. _____

8. _____

9. _____

10. _____

11. _____

12. _____

Chapter 31 Reptiles and Birds, continued

In your textbook, read about the diversity of reptiles and the origins of reptiles.

Complete the chart by checking the correct column(s) for each characteristic.

Characteristic	Snakes	Lizards	Turtles	Crocodiles
13. Guard their nests against predators				
14. Possess shells				
15. Use tongue and Jacobsen's organ for smelling				
16. Kill prey by drowning it				
17. Lack limbs				
18. Have vertebrae and ribs fused to a carapace				
19. Some change color dramatically				
20. Lack teeth				
21. Some inject venom with fangs				
22. Some are aquatic				
23. Are primarily insect eaters				
24. Include marine species that migrate				

Complete each sentence.

25. During the Mesozoic Era, _____ were the most abundant land vertebrates.

26. Snakes and lizards are descended from early _____ _____ , which in

turn were descended from _____ .

27. _____ are probably the modern, living descendants of some type of dinosaur.

Chapter 31 Reptiles and Birds, continued

In your textbook, read about what is a bird.

Answer the following questions.

1. From what type of animal are birds thought to have evolved?

_____

2. List three physical features of birds that link them to reptilian ancestors.

_____

3. Besides making flight possible, what other functions do feathers serve?

_____

4. By what process are old feathers replaced?

_____

In your textbook, read about how birds are adapted for flight.

If the statement is true, write *true*. If it is not, rewrite the italicized part to make it true.

5. A bird's sternum is *the point of attachment for its flight muscles.*

_____

6. Being endothermic, birds have a body temperature that *fluctuates with environmental temperature.*

_____

7. Because of its energy requirements, you might expect a bird to eat *less* than a reptile of comparable size.

_____

8. Hollow bones is an adaptation that makes birds *more efficient predators.*

_____

9. Birds grind up their food in a muscular *gizzard.*

_____

10. The air inside a bird's lungs always has a fairly high *carbon dioxide* content, which makes for efficient gas exchange.

_____

11. For a bird such as a goose or a duck, down feathers are the key to its superior *waterproofing.*

_____

In your textbook, read about the diversity of birds.

For each item in Column A, write the letter of the matching item in Column B.

Column A	Column B
_____ **12.** long beak that is used for dipping into flowers to obtain nectar	**a.** owl
_____ **13.** wings and feet modified for swimming; body surrounded with a thick layer of insulating fat	**b.** pelican
_____ **14.** short, stout beak that is adapted to cracking seeds	**c.** hummingbird
_____ **15.** feathered legs and feet that make it easier to walk in the snow	**d.** penguin
_____ **16.** huge beak with a pouch that is used as a net for capturing fish	**e.** ptarmigan
_____ **17.** large eyes, an acute sense of hearing, and sharp claws; adapted for nocturnal predation	**f.** goldfinch

In your textbook, read about the origins of birds.

Complete each statement.

18. The fossil record of birds is incomplete because bird skeletons are _____ and

_____ .

19. Unlike modern birds, _____ had _____ , a long _____ ,

and _____ _____ _____ .

20. Scientists hypothesize that _____ used its feathers for _____ ,

_____ , or _____ _____ , rather than for flight.

Chapter
32 Mammals

In your textbook, read about what is a mammal, and mammalian hair.

Answer the following questions.

1. Why are mammals able to live in almost every possible environment on Earth?

_____

_____

2. How do sweat glands help regulate body temperature?

_____

_____

Complete the table by checking the column that best fits each example.

	Serves As:		
Example	**Camouflage**	**Defense**	**Warning**
3. The striped fur of a tiger			
4. The sharp quills of a porcupine			
5. A skunk's black-and-white striped fur			
6. The white winter coat of an arctic hare			
7. The white hair patch on a pronghorn			

In your textbook, read about how mammals nurse their young, and about respiration and circulation.

Complete each statement.

8. Female mammals feed their young with milk produced by _____ _____ .

9. In addition to milk and sweat, the glands of mammals produce _____ ,

_____ _____ , and _____ .

10. The milk of mammals is rich in _____ , sugars, _____ , minerals, and

_____ .

11. A mammal's muscular _____ expands the _____ _____ bringing air

into the lungs with each breath.

12. Like birds, mammals have _____ hearts in which _____ _____

is kept entirely separate from _____ _____ .

In your textbook, read about mammalian teeth, limbs, and learning.

Determine if the statement is <u>true</u> or <u>false</u>.

_____ **13.** The size and shape of a mammal's teeth can give valuable clues about its diet.

_____ **14.** Plant-eaters such as horses and cows have well-developed canine teeth for piercing food.

_____ **15.** The teeth of mammals are generally more uniform than the teeth of fishes and reptiles.

_____ **16.** By chewing their cud and then swallowing it, some mammals help bacteria break down the cellulose in their food.

_____ **17.** Mammalian limbs are adapted for a variety of methods of food gathering.

_____ **18.** Moles use their opposable thumbs to grasp objects.

_____ **19.** One reason mammals are successful is that they guard their young and teach them survival skills.

_____ **20.** Complex nervous systems and highly developed brains make it possible for many kinds of mammals to learn.

Circle the letter of the response that best completes the statement.

21. Premolars and molars are used for
  a. shearing. **b.** crushing. **c.** grinding. **d.** all of these.

22. Cud chewing is an adaptation found in
  a. bears and other omnivores. **b.** tigers and other carnivores.
  c. many hoofed mammals. **d.** all of these.

23. The limbs of antelopes are characterized by
  a. greatly elongated finger bones. **b.** strong, slender bones.
  c. short bones and large claws. **d.** none of these.

24. Chimpanzees are intelligent enough to
  a. use tools. **b.** use sign language.
  c. work machines. **d.** all of these.

Chapter 32 Mammals, *continued*

Section 32.2 Diversity of Mammals

In your textbook, read about placental mammals, mammals with a pouch, and egg-laying mammals.

Answer the following questions.

1. What is a placental mammal?

_____

_____

2. What is the relationship between body size and gestation period in placental mammals?

_____

3. Why are most marsupials found only in and around Australia?

_____

_____

_____

4. What characteristic sets monotremes apart from all other mammals?

_____

_____

_____

Complete the table by checking the correct column(s) for each characteristic.

Characteristic	Type of Mammal		
	Placental	Marsupial	Monotreme
5. Give birth to young			
6. Young nourished by a placenta during the entire development period			
7. Have a permanent pouch on abdomen			
8. Produce milk in mammary glands			
9. Lay eggs			
10. Include echidnas and duck-billed platypuses			
11. Have hair			
12. Comprise about 95 percent of all mammals			
13. Exhibit parental care			

Chapter 32   **Mammals,** continued

In your textbook, read about the origins of mammals.

Use each of the terms below just once to complete the passage.

climate	dinosaurs	insect-eating
mammals	Pangaea	reptiles therapsids

Roughly 200 million years ago, the first **(14)** _____ began roaming Earth. Most were small **(15)** _____ descendants of **(16)** _____ , animals that had characteristics in common with both mammals and **(17)** _____ . The breakup of **(18)** _____ , the sudden disappearance of the **(19)** _____ , and a changing **(20)** _____ provided early mammals with new food sources and habitats.

Examine the phylogenetic tree below. Then answer the question that follows.

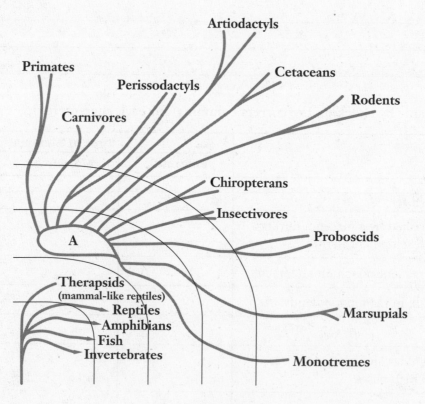

21. What group of animals is represented by the letter A above? _____

Reinforcement and Study Guide

Chapter 33 Animal Behavior

Section 33.1 Innate Behavior

In your textbook, read about what behavior is, inherited behavior, automatic responses to stimuli, and instinctive behavior.

Answer the following questions.

1. What is meant by animal behavior?

_____

_____

2. How is behavior adaptive?

_____

_____

3. Explain the relationship between innate behaviors and genetics.

_____

_____

_____

4. What is an instinct?

_____

Identify each of the following as being either a reflex or an instinct.

_____ **5.** You leap up after sitting down in shorts on a hot car seat.

_____ **6.** A sea turtle returns to the beach where she was hatched, in order to lay her eggs.

_____ **7.** A giant clam closes its shell when a shadow falls across it.

_____ **8.** A spider spins a complex, circular web.

In your textbook, read about courtship behavior and territoriality.

Determine if the following statements are <u>true</u> or <u>false</u>.

_____ **9.** Courtship behavior is something only male animals can instinctively perform.

_____ **10.** Courtship behavior is adaptive because it ensures that members of the same species can recognize each other and mate.

_____ **11.** A territory is a physical space that one animal defends against all other species of animals.

_____ **12.** Setting up territories reduces conflicts between members of the same species.

Chapter 33 **Animal Behavior,** continued

In your textbook, read about aggressive behavior, submission, and behavior resulting from internal and external cues.

Below are excerpts from the field notebook of a behavorial biologist. Identify the behavioral phenomenon being described in each case.

13.

> **Field Notes**
>
> A large male baboon stares at another male and then suddenly "yawns" to reveal his long, sharp fangs.
>
> **Behavior Exhibited:**
>
> _____

14.

> **Field Notes**
>
> When a herd of elephants arrives at a waterhole, the oldest female drinks first, followed by three females with calves, and finally a young male.
>
> **Behavior Exhibited:**
>
> _____

15.

> **Field Notes**
>
> After fighting briefly with an older pack member, a young wolf stops fighting and rolls onto her back with her tail tucked between her legs and her eyes averted.
>
> **Behavior Exhibited:**
>
> _____

16.

> **Field Notes**
>
> Large numbers of monarch butterflies fly south to roost in the winter.
>
> **Behavior Exhibited:**
>
> _____

Complete the following sentences.

17. The type of dominance hierarchy formed by chickens is called a(n) _____ _____ .

18. A cycle of behavior that occurs roughly every 24 hours is known as a(n) _____

_____ .

19. Some animals use the positions of the _____ _____ _____

to navigate. Others may use _____ clues or Earth's _____

_____ .

20. _____ is similar to hibernation, in that metabolic activity _____

in response to internal and external cues.

In your textbook, read about learned behavior.

Answer the following questions.

1. What is learned behavior?

   _____

   _____

2. What is a major advantage of being able to learn?

   _____

   _____

   _____

Complete the table by checking the correct column for each example.

	Type of Behavior	
Example	**Learned**	**Innate**
3. A dog catching a disk		
4. A dog scent-marking a tree with urine		
5. A parrot saying "Polly want a cracker"		
6. A young lioness stalking prey with her mother		
7. A woodchuck going underground to hibernate		

In your textbook, read about habituation, imprinting, and learning by trial and error.

For each item in Column A, write the letter of the matching item in Column B.

Column A	Column B
_____ **8.** You stay with relatives who have a clock that chimes every hour. The first two nights, the chimes keep you awake, but after that you no longer notice them.	**a.** imprinting
_____ **9.** A boy receives a day-old duckling as gift. It soon follows the boy wherever he goes.	**b.** trial-and-error learning
_____ **10.** A young woman takes up archery. At first, her arrows don't hit the target, but after a week of practice, she is hitting the bull's eye fifty percent of the time.	**c.** habituation

In your textbook, read about conditioning and insight.

If the statement is true, write *true*. If it is not, rewrite the italicized part to make it true.

11. Learning by *repeating something over and over* is known as conditioning. _____

12. In Pavlov's conditioning experiments, the *innate reflex of salivating* was the stimulus that the dogs
learned to associate with food. _____

13. Once conditioned, Pavlov's dogs would salivate at the sound of the bell *even when no food was present.*

_____

14. A child figuring out how to use a chair to reach a cookie jar is an example of *trial and error.*

_____

15. Insight is learning in which an animal uses *previous experience* to respond to a new situation.

_____

In your textbook, read about the role of communication.

Use each of the terms below just once to complete the passage.

behavior	communication	information	innate	language
meanings	odors	pheromones	sounds	symbols

   Through various forms of **(16)** _____ , animals exchange **(17)** _____

that affects their **(18)** _____ . Animals can communicate with visual signals, by touching

each other, and by producing **(19)** _____ , some of which can be heard over great dis-

tances. Communicating with **(20)** _____ is another strategy; **(21)** _____ are

species-specific odor chemicals that can have a powerful effect on behavior. Some types of communication

involve both **(22)** _____ and learned behavior. Human **(23)** _____ has

evolved as a way of communicating with written and spoken **(24)** _____ that have specific

(25) _____ .

BioDigest 9 Vertebrates

In your textbook, read about fishes, amphibians, reptiles, birds, and mammals.

FISHES

Complete the chart by checking the correct column(s) for each characteristic of fishes.

Adaptation	Jawless	Cartilaginous	Bony
1. Jaws			
2. Gills			
3. Lateral line system			
4. Paired fins			
5. Skeleton made of cartilage			
6. Swim bladder			

AMPHIBIANS

Complete the following sentences.

7. Amphibians are _____ , animals that have variable body temperatures and obtain heat from external sources. These vertebrates also carry out gas exchange through their _____ . Most amphibians live on _____ but reproduce in _____ . Almost all amphibians go through _____ , a radical change between the form of the young and the form of the adult.

REPTILES

Complete the table by describing the advantages that reptiles have because of certain adaptations. List one advantage for each adaptation.

Adaptation	Advantage
8. Scaly skin	
9. Amniotic egg	

BIRDS

Flight affects almost every system in birds. Explain the flight adaptations in each system listed here.

System	Adaptation
10. Bones	
11. Respiration	
12. Body covering	
13. Legs	
14. Wings	

MAMMALS

Various adaptations of mammals serve certain functions. In the space provided, write the letter of the adaptations that perform the function. Any letter may be used more than once.

Function	Adaptation
15. protection from low temperatures	**a.** hair
	b. sweat glands
16. protection from high temperatures	**c.** four-chambered heart
	d. diaphragm
17. feeding young	**e.** canine teeth
	f. mammary glands
18. stabbing or holding food	**g.** small ears
	h. body fat
19. grinding or chewing food	**i.** molars and premolars
	j. hibernation
20. providing large amounts of oxygen	**k.** estivation

REINFORCEMENT AND STUDY GUIDE

Reinforcement and Study Guide

Section 34.1 Skin: The Body's Protection

In your textbook, read about the structure and function of the skin.

Complete the table by checking the correct column for each description.

Description	Epidermis	Dermis
1. The outermost layer of skin		
2. Contains connective tissue, glands, and muscles		
3. The thicker, inner layer of skin		
4. Partly composed of dead, keratin-containing cells		
5. Contains pigmented cells that protect against the sun's rays		
6. Hair follicles grow out of this layer		
7. Site of continual mitotic cell divisions		
8. Richly supplied with blood vessels and nerves		

Answer the following questions.

9. Describe the change that takes place in your skin when you are exposed to ultraviolet light.

_____

_____

_____

_____

10. How does skin help regulate body temperature?

_____

_____

_____

_____

11. List three other functions of skin.

_____

_____

_____

_____

In your textbook, read about the structure of the skeletal system and joints.

Identify the following as being part of the axial or appendicular skeleton.

_____ **1.** the tarsals, metatarsals, and phalanges in your foot

_____ **4.** the bones in your shoulder

_____ **2.** the seven vertebrae in your neck

_____ **5.** your lower jaw

_____ **3.** your rib cage

_____ **6.** the humerus in your arm

For each answer below, write an appropriate question.

7. **Answer:** They are bands of connective tissue that attach muscles to bones.

Question: _____

8. **Answer:** Fluid-filled sacs that reduce friction between bones in a joint.

Question: _____

9. **Answer:** They connect bones to other bones.

Question: _____

10. **Answer:** One allows the bones to move back and forth; the other allows the bones to rotate.

Question: _____

In your textbook, read about the formation of bone and bone growth.

Complete each sentence.

11. In a human embryo's skeleton, _____ is gradually replaced by _____ except in a few places like the tip of the _____ .

12. Some cells in cartilage are stimulated to become _____ . They secrete a substance in which _____ _____ and other minerals are deposited.

13. Your bones increase in length near their _____ .

14. Even after you reach your full adult height, the bone-forming cells in your body will still be involved in _____ and _____ .

In your textbook, read about compact and spongy bone and skeletal system functions.

Answer the following questions.

15. If you cut through to the center of a large leg bone, what bone components (in order, from the outside in) would you encounter?

_____

_____

16. How do blood vessels and nerves reach individual bone cells in compact bone?

_____

17. What role does bone marrow play in the functioning of your circulatory system?

_____

_____

18. In what way is the skeleton a storehouse?

_____

_____

_____

In your textbook, read about growth, mineral storage, and injury and disease in bone.

Determine if the statement is <u>true</u> or <u>false</u>.

_____ **19.** Once you have finished growing, your bones no longer change.

_____ **20.** Calcium is both deposited in and removed from bones.

_____ **21.** Calcium removed from bone is rapidly excreted in the urine as an unnecessary body waste.

_____ **22.** As a person ages, his or her bone density usually decreases.

_____ **23.** Because bones in an adult's skeleton are harder than children's bones, adults are less likely to break a bone in a fall.

_____ **24.** Osteoporosis is most common in older women because they rarely include milk in their diet.

In your textbook, read about three types of muscles and skeletal muscle contraction.

Complete the table by checking the correct column for each description.

Description	Type of Muscle		
	Smooth	Skeletal	Cardiac
1. under voluntary control			
2. striated			
3. slow, prolonged contractions			
4. attached to bones			
5. found only in the heart			
6. not under voluntary control			
7. lines cavities and surrounds organs			

In your textbook, read about muscle strength and exercise.

If the statement is true, write <u>true</u>. If it is not, rewrite the italicized part to make it true.

8. Muscle strength depends on *the number of fibers in a muscle*.

_____

9. When oxygen is limited, *aerobic respiration* becomes a muscle's primary source for ATP.

_____

10. During lactic acid fermentation, *oxygen* builds up in muscle cells.

_____

11. A drop in the amount of lactic acid in the bloodstream indicates that muscular activity has *decreased*.

_____

Chapter 35 The Digestive and Endocrine Systems

In your textbook, read about the functions of the digestive tract, the mouth, and the stomach.

Complete each statement.

1. The entire process of digestion involves first _____ food, then _____ it into simpler compounds, then _____ nutrients for use by body cells, and, finally, _____ wastes.

2. By chewing your food, you _____ its surface area.

3. Various enzymes play a role in _____ digestion, while the action of teeth, tongue, and muscles are involved in _____ digestion.

4. In your mouth, the enzyme _____ is released from _____ glands to begin the chemical breakdown of _____ .

5. Your _____ are adapted for cutting food, while your _____ are best suited for grinding food.

If the statement is true, write *true*. If it is not, rewrite the italicized part to make it true.

6. During swallowing, the epiglottis covers the *esophagus* to prevent choking.

_____

7. Food is moved through the digestive tract by rhythmic waves of *voluntary muscle contractions* called peristalsis.

_____

8. The churning actions of the stomach help mix the food with *pancreatic juices*.

_____

9. Pepsin is a *protein-digesting enzyme* that only works in an acidic environment.

_____

10. The stomach releases its contents into the small intestine *suddenly, all at once.*

_____

In your textbook, read about the small intestine and the large intestine.

Answer the following questions.

11. What role do the enzymes secreted by the pancreas play in the digestive process?

_____

_____

_____

12. Explain the relationship between the liver, the gallbladder, and bile.

_____

_____

_____

13. Once in the small intestine, what happens to

a. digested food?

_____

_____

b. indigestible materials?

_____

_____

Complete the table by checking the correct column(s) for each function.

Function	Small Intestine	Large Intestine
14. Water is absorbed through walls.		
15. Digestion is essentially completed.		
16. Vitamin K is produced.		
17. Nutrients are absorbed by villi.		
18. Contents are moved by peristalsis.		
19. Indigestible material is collected.		
20. Bile and pancreatic juices are added.		

In your textbook, read about carbohydrates, fats, and proteins.

Complete the table by checking the correct column(s) for each description.

Description	Carbohydrates	Fats	Proteins
1. the most energy-rich nutrients			
2. sugars, starches, and cellulose			
3. broken down into amino acids			
4. part of a nutritious, balanced diet			
5. normally used for building muscle, but can be used for energy			
6. broken down into glucose, fructose, and other simple sugars			
7. used to insulate the body from cold			

In your textbook, read about minerals and vitamins, water, and metabolism and calories.

Complete each statement.

8. _____ are inorganic substances that help to build tissue or take part in chemical reactions in the body.

9. Unlike minerals, _____ are organic nutrients that help to regulate body processes.

10. The two major vitamin groups are the _____ and the _____ vitamins.

11. The energy content of food is measured in _____ , each of which is equal to _____ calories.

12. Despite the claims of many fad diets, the only way to lose weight is to _____ more calories than you _____ .

In your textbook, read about control of the body and negative feedback control.

Complete each statement.

1. Internal control of the body is handled by the _____ system and the

   _____ system.

2. Most endocrine glands are controlled by the action of the _____ ,

   or master gland.

3. A(n) _____ is a chemical released in one part of the body that affects

   another part.

4. The amount of hormone released by an endocrine gland is determined by the body's

   _____ for that hormone at a given time.

5. A _____ system is one in which hormones are fed back to inhibit the

   original signal.

6. When your body is dehydrated, the pituitary releases ADH hormone, which reduces the amount

   of _____ in your urine.

7. When you have just eaten and your blood glucose levels are high, your pancreas releases the

   hormone _____ , which signals the liver to take in glucose, thereby

   lowering blood glucose levels.

In your textbook, read about hormone action, adrenal hormones and stress, and other hormones.

For each item in column A, write the letter of the matching item from Column B.

Column A	Column B
_____ 8. Determines the body's food intake requirements	**a.** steroid hormones
_____ 9. Made from lipids and diffuse freely into cells through the plasma membrane	**b.** glucocorticoids and aldosterone
_____ 10. Bind to receptors embedded in the plasma membrane of the target cell.	**c.** calcitonin and parathyroid hormone
_____ 11. Produce a feeling called "adrenaline rush"	**d.** epinephrine and norepinephrine
_____ 12. Help the body prepare for stressful situations	**e.** amino acid hormones
_____ 13. Regulate calcium levels in blood	**f.** thyroxine

Chapter 36 The Nervous System

In your textbook, read about neurons–basic units of the nervous system.

Complete the table by filling in the missing information in each case.

Structure	Function
1.	carry impulses toward the brain and spinal cord
2. dendrites	
3. motor neurons	
4.	transmit impulses within the brain and spinal cord
5.	carry impulses away from neuron cell bodies

Order the steps in impulse transmission from 1 to 7.

_____   **6.** A wave of depolarization moves down the neuron.

_____   **7.** The Na^+/K^+ pump takes over again, pumping sodium ions out across the membrane, and pumping potassium ions in.

_____   **8.** Sodium channels in the neural membrane open.

_____   **9.** A neuron receives a stimulus.

_____   **10.** As the wave of depolarization passes, sodium channels close and potassium channels open.

_____   **11.** The neuron returns to a resting state.

_____   **12.** Sodium ions flow into the neuron, causing the inside of the neuron to become positively charged.

Chapter 36 The Nervous System, continued

In your textbook, read about the central nervous system and the peripheral nervous system.

Label the diagram of the brain to show the cerebrum, cerebellum, and brain stem.

13. _____

14. _____

15. _____

Write the name of the part labeled above that matches each description in the table.

Description	Part
16. Includes the medulla and pons	
17. Controls conscious activities and movement	
18. Important for keeping your balance	
19. If damaged, heart rate might be affected	
20. If damaged, memory might be affected	
21. Ensures that movements are coordinated	

Complete the table by checking the correct column for each description.

	Autonomic Nervous System Division	
Description	**Sympathetic**	**Parasympathetic**
22. Controls internal activities when the body is at rest		
23. Increases breathing rate		
24. Tenses muscles		
25. Slows heart rate down		
26. Activates fight or flight response		

In your textbook, read about sensing chemicals and sensing light.

Determine if each statement is <u>true</u> or <u>false</u>.

_____ **1.** Impulses coming from sensory receptors in your nose and mouth are interpreted as odors and tastes by the cerebrum.

_____ **2.** The senses of taste and smell are closely linked.

_____ **3.** The lens in the eye controls the amount of light that strikes the retina.

_____ **4.** On a bright sunny day, the cones in your eyes play a greater role in your sense of sight than the rods.

_____ **5.** Only the left hemisphere of the brain is involved in the sense of sight.

_____ **6.** When you are looking at an object, each of your eyes sees the object from the same perspective.

_____ **7.** The retina contains two types of light receptor cells.

In your textbook, read about sensing mechanical stimulation.

Circle the letter of the response that best completes each statement.

8. Sound waves are converted into nerve impulses inside the
  a. ear canal.   **b.** cochlea.   **c.** malleus.   **d.** optic nerve.

9. If the semicircular canals in one of your ears were damaged, you might
  a. lose your ability to hear low-frequency sounds.
  b. lose your ability to coordinate your neck muscles.
  c. lose your sense of balance.
  d. lose your sense of rhythm.

10. The malleus, incus, and stapes are found in the
  a. outer ear.   **b.** eardrum.   **c.** middle ear.   **d.** inner ear.

11. Your senses of hearing and touch *both* depend on nerve impulses being generated by
  a. electrical stimulation.   **b.** sound waves.
  c. a change in temperature.   **d.** mechanical stimulation.

12. In the skin of your fingertips, you might expect to find receptors for
  a. touch.   **b.** pressure.   **c.** pain.   **d.** all of these

Chapter 36 **The Nervous System,** *continued*

In your textbook, read about how drugs act on the body, their medicinal uses, and abuse of drugs.

Answer the following questions.

1. Distinguish between a drug and a medicine.

_____

_____

_____

2. What is a narcotic?

_____

3. Compare the effect of a stimulant on the CNS with the effect of a depressant.

_____

_____

4. What is an addiction?

_____

5. How does a person's body develop a tolerance for a drug?

_____

_____

_____

In your textbook, read about the classes of commonly abused drugs.

Complete the table by checking the correct column for each example.

Example	Stimulant	Depressant
6. Drugs that cause an increase in heart rate		
7. Alcohol		
8. Nicotine		
9. Caffeine		
10. Barbiturates		
11. Drugs that cause vasoconstriction		
12. Opiates		
13. Hallucinogens		

Chapter 37 Respiration, Circulation, and Excretion

In your textbook, read about air passageways and lungs.

Circle the letter of the choice that best completes the statement or answers the question.

1. During the process of respiration,
   a. oxygen is delivered to body cells.
   b. carbon dioxide is expelled from the body.
   c. oxygen is used in cells to produce ATP.
   d. all of these.

2. When you swallow, your epiglottis momentarily covers the top of the trachea so that
   a. you can swallow more easily.
   b. you can breathe more easily.
   c. you don't get food in your air passages.
   d. you can cough up foreign matter.

3. The cilia that line your trachea and bronchi
   a. produce dirt-trapping mucus.
   b. help in the exchange of oxygen and CO_2.
   c. move mucus and dirt upward.
   d. only beat when you inhale.

4. The first branches off the trachea are called
   a. bronchioles.
   b. bronchi.
   c. arterioles.
   d. alveoli.

5. Inside the alveoli, carbon dioxide and oxygen
   a. are exchanged between air and blood.
   b. are transported along microscopic tubules.
   c. are produced inside cells.
   d. are exchanged for other gases.

6. Which is the correct sequence for the path of oxygen through the respiratory system?
   a. nasal passages, bronchi, trachea, bronchioles, cells, blood, alveoli
   b. cells, blood, alveoli, bronchioles, bronchi, trachea, nasal passages
   c. nasal passages, blood, alveoli, bronchi, cells, trachea, bronchioles
   d. nasal passages, trachea, bronchi, bronchioles, alveoli, blood, cells

In your textbook, read about the mechanics of breathing and the control of respiration.

For each statement below, write <u>true</u> or <u>false</u>.

_____ 7. Homeostasis in respiration is controlled by the cerebrum.

_____ 8. As you exhale, the bronchioles in the lungs release most of their air.

_____ 9. When you inhale, the muscles between your ribs contract.

_____ 10. Relaxation of the diaphragm causes a slight vacuum in the lungs.

_____ 11. Air rushes into the lungs because the air pressure outside the body is greater than the air pressure inside the lungs.

_____ 12. Relaxation of the diaphragm causes it to flatten.

Reinforcement and Study Guide

Section 37.2 The Circulatory System

In your textbook, read about your blood, ABO blood types, and blood vessels.

Answer the following questions.

1. What cells and substances would you expect to find suspended or dissolved in plasma?

_____

_____

2. How is carbon dioxide transported in blood?

_____

_____

Complete the table below by checking the correct column for each description.

Description	Red Blood Cells	White Blood Cells	Platelets
3. Contain hemoglobin			
4. Fight infection			
5. Lack a nucleus			
6. Help clot blood			
7. Transport oxygen			
8. Comparatively large and nucleated			

For each statement below, write <u>true</u> or <u>false</u>.

_____ **9.** Your blood type can be changed with a blood transfusion.

_____ **10.** Different blood types result from different antibodies being present on the membranes of red blood cells.

_____ **11.** If you have type B blood, then you have anti-A antibodies in your plasma.

_____ **12.** Risks involving incompatible Rh factors are greatest for a woman's first child.

In your textbook, read about your heart, blood's path through the heart, and inside your heart.

Label the parts of the human heart in the diagram below. Use these choices:

aorta	left atrium	left ventricle	pulmonary arteries
pulmonary veins	right atrium	right ventricle	

13. _____

14. _____

17. _____

19. _____

15. _____

16. _____

18. _____

20. Where does blood go from the pulmonary veins? From the right ventricle? From the left ventricle?
_____

21. What prevents blood from mixing between atria and ventricles?
_____

In your textbook, read about heartbeat regulation, control of the heart, and blood pressure.

Determine if the statement is true. If it is not, rewrite the italicized part to make it true.

22. The surge of blood through an artery is called the *cardiac output.* _____

23. The pacemaker initiates heartbeats *by generating electrical impulses.* _____

24. An electrocardiogram is a record of *the strength of each heartbeat.* _____
_____

25. The *atrioventricular node*, along with sensory cells in arteries near the heart, regulates the pacemaker.
_____

26. *Diastolic pressure* occurs when the heart's ventricles contract. _____

Chapter 37 Respiration, Circulation, and Excretion, *continued*

In your textbook, read about kidneys, nephrons, and the formation of urine.

Answer the following questions.

1. What is the major function of kidneys?

_____

2. What role does the bladder play in the urinary system?

_____

3. What are nephrons?

_____

Order the following steps in the filtration of blood from 1 to 7.

_____ **4.** From the Bowman's capsule, fluid flows through a U-shaped tubule.

_____ **5.** Under high pressure, blood flows into capillaries that make up the glomerulus.

_____ **6.** After being stored in the bladder, urine exits the body via the urethra.

_____ **7.** Fluid moves from the end of the nephron's tubule to the ureter.

_____ **8.** Blood enters the nephron from a branch of the renal artery.

_____ **9.** Water, glucose, amino acids, and ions are reabsorbed into the blood.

_____ **10.** Water, glucose, amino acids, wastes, and other substances move from glomerular capillaries into a Bowman's capsule.

In your textbook, read about the urinary system and homeostasis.

Complete each statement.

11. _____ and _____ are two toxic nitrogenous wastes

that your kidneys constantly remove from your bloodstream.

12. The kidneys also help regulate the blood's _____ _____ ,

and _____ .

13. Individuals with diabetes have excess levels of _____ in their blood.

In your textbook, read about human male anatomy and hormonal control.

Answer the following questions.

1. What are the primary functions of the male reproductive system?

_____

_____

2. How does the location of the scrotum affect sperm?

_____

_____

3. How many sperm can the average mature male produce in one day?

_____

Order the steps in the formation and transportation of sperm from 1 to 6.

_____ **4.** Mature sperm enter the vas deferens.

_____ **5.** Newly formed haploid sperm cells pass through a series of coiled ducts to the epididymis.

_____ **6.** Sperm leave the body via the urethra.

_____ **7.** Sperm mature in the epididymis.

_____ **8.** Cells lining tubules in the testes undergo meiosis.

_____ **9.** Sperm travel along the ejaculatory ducts and into the urethra.

Complete each sentence.

10. When a young man's voice "changes," he is probably entering _____ , a time when he will develop other secondary _____ _____ .

11. A hormone released by the _____ stimulates the _____ gland to release _____ - _____ and _____ hormones.

12. FSH regulates _____ production, while LH controls the production of the steroid hormone _____ by the testes.

Chapter 38 Reproduction and Development, continued

In your textbook, read about human female anatomy and puberty in females.

If the statement is true, write <u>true</u>. If it is not, rewrite the italicized part to make it true.

13. When an egg cell is released from an ovary, it moves down the oviduct by *gravity*.

_____

14. As is the case in human males, a woman's *hypothalamus* produces FSH and LH.

_____

15. *FSH* stimulates follicle development and the release of estrogen from the ovary.

_____

16. In females, *luteinizing hormone (LH)* is responsible for the development of the secondary sex characteristics.

_____

17. Long before a woman is born, cells in her ovaries that are destined to become future eggs undergo several *mitotic divisions*.

_____

In your textbook, read about the menstrual cycle.

Complete the table by checking the correct column for each event.

Event	Phase of Menstrual Cycle		
	Flow	Follicular	Luteal
18. LH stimulates the corpus luteum to develop from a ruptured follicle.			
19. Estrogen levels are at their peak.			
20. A cell inside a follicle resumes meiotic divisions.			
21. Progesterone levels are at their peak.			
22. The uterine lining is shed.			
23. LH levels rise abruptly.			
24. Ovulation occurs.			
25. The uterine lining becomes engorged with blood, fat, and tissue fluid.			
26. FSH begins to rise.			

In your textbook, read about fertilization and implantation, and embryonic membranes.

Use each of the terms below just once to complete the passage.

amnion	blastocyst	chorion	chorionic villi	embryo
implants	oviduct	placenta	umbilical cord	zygote

Usually in the upper part of a(n) **(1)** _____ , an egg and one sperm unite

to form a(n) **(2)** _____ . This single cell divides repeatedly to form a(n)

(3) _____ , which **(4)** _____ in the uterine wall. Part

of the blastocyst becomes the **(5)** _____ , which is surrounded by a fluid-filled,

membranous sac called the **(6)** _____ . The embryo is connected to the wall

of the uterus by its **(7)** _____ . The amniotic sac is enclosed by the

(8) _____ , which later forms the **(9)** _____ .

Nutrients and oxygen from the mother and wastes from the embryo are exchanged in the

(10) _____ .

In your textbook, read about fetal development and genetic counseling.

Complete the table by checking the correct column for each event or example.

	Trimester		
Event/Example	**First**	**Second**	**Third**
11. Fetus can survive outside the uterus with medical assistance.			
12. Fetus weighs more than 3000 grams.			
13. Embryo is most vulnerable to outside influences.			
14. Embryo becomes a fetus.			
15. Fetus can use its muscles to move spontaneously.			
16. Fetus becomes oriented head-down.			
17. Gender of fetus can be determined.			

In your textbook, read about birth.

Answer the following questions.

1. What are the three stages of birth? _____

2. Describe the action of oxytocin. _____
_____

3. After the placenta is expelled from a woman's body, what effect do continued uterine contractions
have? _____
_____

In your textbook, read about growth and aging.

Complete each sentence.

4. Your growth rate, as well as the type of growth you undergo, varies with both your
_____ and your _____ .

5. _____ _____ _____ ,
abbreviated _____ , regulates growth.

6. hGH exerts its effects primarily on _____ and on _____
_____ .

7. LGH works by increasing _____ and _____ .

Complete the table by checking the correct column for each description.

Example	Childhood	Adolescence	Adulthood
8. Your growth rate continues at a steady rate.			
9. Lines develop on your face, especially around your eyes and mouth.			
10. You reach maximum physical stature.			
11. You begin to reason.			
12. You may have a sudden growth spurt.			

Chapter 39 **Immunity from Disease**

Section 39.1 The Nature of Disease

In your textbook, read about what an infectious disease is, determining what causes a disease, and the spread of infectious diseases.

Answer the following questions.

1. Why is a disease like osteoarthritis not considered an infectious disease?

_____

2. What is meant by Koch's postulates?

_____

_____

3. In terms of disease, what is a reservoir?

_____

Complete the table by writing in the method of transmission for each example.

Example	Method of Transmission
4. While exploring a cave, a person breathes in fungal spores that cause a lung infection.	
5. A person contracts Rocky Mountain spotted fever after being bitten by a tick.	
6. After having unprotected sex, a person contracts syphilis.	

In your textbook, read about what causes the symptoms of a disease, patterns of disease, and treating diseases.

For each statement below, write _true_ or _false_.

_____ 7. The toxin produced by a particular microorganism might be far more destructive than the direct damage the microbe does to its host cells.

_____ 8. Endemic diseases often disappear in a population, only to resurface unexpectedly many years later.

_____ 9. If you catch the flu during an influenza epidemic, your best hope of recovery is to take antibiotics.

_____ 10. It is important for researchers to try to discover new antibiotics because many types of bacteria are becoming resistant to the ones now being used.

Chapter 39 Immunity from Disease, continued

Section 39.2 Defense Against Infectious Diseases

In your textbook, read about the innate immune system.

If the statement is true, write *true*. If it is not, rewrite the italicized part to make it true.

1. Healthy skin is a good defense against the invasion of pathogens because it is *free of bacteria*.

_____

_____

2. In your trachea, *saliva* traps microbes and prevents them from entering your lungs.

_____

3. Macrophages migrate *into the bloodstream* when the body is challenged by a pathogen.

_____

4. Phagocytes at the site of an infection or inflammation destroy pathogens by surrounding and engulfing them. _____

5. The third method of defense against infection is the consumption of pathogens by *neutrophils*.

_____

6. Interferon is produced by cells infected by *pathogenic bacteria*.

_____

In your textbook, read about acquired immunity.

Circle the letter of the choice that best completes the statement.

7. The human lymphatic system is important in
   a. filtering pathogens from lymph.
   c. resistance to disease.
   b. keeping body fluids constant.
   d. all of the above.

8. Tissue fluid is found
   a. in lymph vessels.
   c. around body cells.
   b. in the bloodstream.
   d. in lymph ducts.

9. The main function of lymph nodes is to
   a. store red blood cells.
   c. filter excess fluid.
   b. filter lymph.
   d. trigger an immune response.

10. A reservoir for lymphocytes that can be transformed into specific disease-fighting cells is the
   a. thymus gland.
   c. pituitary gland.
   b. thyroid gland.
   d. pancreas.

REINFORCEMENT AND STUDY GUIDE

In your textbook, read about antibody immunity and cellular immunity.

Complete each sentence.

11. _____ is the building up of a _____ to a specific pathogen.

12. Two types of immunity that involve different kinds of cells and cellular actions are _____ immunity and _____ immunity.

13. The presence of foreign _____ in the body triggers the production of _____ by plasma cells.

14. A _____ is a lymphocyte that, when activated by a _____, becomes a plasma cell and produces _____ .

15. Cellular immunity involves several different types of _____ cells.

16. A _____ releases enzymes directly into the _____ .

Complete the table by checking the correct columns for each example.

	Type of Immunity	
Example	**Cellular**	**Antibody**
17. Involves the protection of antibodies		
18. Simulated by antigens in the body		
19. Clones of killer T cells produced		
20. Memory cells produced so the body can respond quickly to a second attack		
21. Key role played by antigen-antibody complex		
22. T cells destroyed by pathogens directly		

Section 39.2 Defense Against Infectious Diseases

In your textbook, read about passive and active immunity to infectious diseases.

Answer the following questions.

23. Distinguish between active and passive immunity.

_____

_____

_____

24. In what two ways can passive immunity develop?

_____

_____

25. What is a vaccine?

_____

_____

In your textbook, read about AIDS and the immune system.

For each statement below, write <u>true</u> or <u>false</u>.

_____ **26.** The virus that causes AIDS—Human Immunodeficiency Virus—is well-named because it attacks the immune system.

_____ **27.** HIV can be transmitted by air.

_____ **28.** A child born to a woman who is infected with HIV is at risk for being infected, too.

_____ **29.** HIV destroys a person's resistance to disease by attacking and destroying memory T cells.

_____ **30.** In a blood sample from an HIV-positive person, you would expect to find most of the viruses existing free in the blood, rather than being hidden inside cells.

_____ **31.** If a person is infected with HIV, he or she will usually develop AIDS within about a year.

_____ **32.** The cause of death for a person with AIDS usually is some type of infection that the body's weakened immune system can no longer fight off.

_____ **33.** The majority of untreated persons infected with HIV will develop AIDS.

BioDigest 10 The Human Body

In your textbook, read about skin, bones, and muscles.

Skin has four functions: **(1)** _____ , **(2)** _____

_____ , **(3)** _____ , and **(4)** _____

_____ . These functions help maintain homeostasis in the body.

Complete the table to describe the role of bones.

Support for	**5.**
Place for	**6.**
Protects	**7.**
Manufacture of	**8.**
Storehouse of	**9.**

This diagram shows the various steps involved in the respiratory process. In the space provided, describe the steps as indicated.

Step 1: Oxygen enters the lungs when you inhale.

Step 7: Carbon dioxide leaves the lungs when you exhale.

LUNG

Step 2: **(10)** _____

_____

_____

Step 6: **(13)** _____

_____

_____

Bloodstream

Bloodstream

Step 3: Oxygen passes from blood to the cells.

Cell

O_2 + Glucose

ATP

CO_2

Step 5: **(12)** _____

_____

_____

_____

Step 4: **(11)** _____

_____

_____

BioDigest
10 **The Human Body,** *continued*

In your textbook, read about reproductive, endocrine, and lymphatic systems.

14. The endocrine system is a communication system; its messages are hormones. They are produced by _____ and travel in the _____ to _____ _____ . There they _____ .

15. The structures of the male reproductive system—the scrotum, _____ , epididymis, seminal vesicles, _____ , bulbourethral gland, urethra, and _____ —are involved in _____ and maintaining sperm cells and _____ into the female reproductive tract.

16. The structures of the female reproductive system—the _____ , oviduct, _____ , and vagina—produce and maintain _____ , receive and transport _____ , and support the development of the _____ .

Explain how these systems of the body interact with each other.

17. Skeletal system ⎯⎯⎯⟶ Circulatory system ⎯⎯⎯⟶ Muscular system

_____

_____

18. Digestive system ⎯⎯⎯⟶ Circulatory system ⎯⎯⎯⟶ Urinary system

_____

_____

19. Endocrine system ⎯⎯⟶ Reproductive system

_____

_____

20. Respiratory System ⎯⎯⎯⟶ Circulatory system ⎯⎯⎯⟶ Cell ⎯⎯⎯⟶ Circulatory system ⎯⎯⎯⟶ Urinary and Respiratory systems

_____

_____

_____

21. Circulatory system ⎯⎯⟶ Lymphatic system

_____

_____

Study Skills

A. Improve Your Reading Skills

Active readers are good readers.

Active readers

- get ready before they read.
- use skills that help them when they read.
- review to remember after they read.

Here's what you can do to become an active reader!

Before You Read

Get Ready to Read

- Find a quiet time and place to read—library, study hall, home.
- Don't read when you're tired.
- Don't read when you're hungry.
- Wait until you have finished a section before you take a break.

Scan

- Quickly scan the material so you will know what it is about.
- Look at pictures and read the captions, titles, headings, and words in bold print.

Write

- Write notes about what you see when you scan.
- Write questions about what you see.
- Write topics you want to find out about when you read.
- Write a preview outline from the section topics.

As You Read

- Find the main idea of each section or paragraph—this is usually in the first sentence.
- Study the pictures, maps, graphs, and tables, and think about the information in them.
- Write down the main ideas and other notes about what you read.
- After you read the whole section, reread the parts you didn't understand.

Study Skills

After You Read

- Review your outline or the notes you wrote while you were reading.
- If you still have questions, ask a classmate or your teacher for help.
- Write important facts or ideas on flash cards.
- Review your flash cards to help you remember what you've read.

B. Improve Your Vocabulary Skills

Active readers learn the meanings of new words.

Active readers

- recognize clues to help find the meaning.
- look for familiar words and word parts in new words.
- use a dictionary often.
- practice new words so they can remember new meanings.

Here's how you can improve your vocabulary!

When You See a New Word

Scan

- Read the sentence and look for clues about the meaning of the word. These are called context clues.
- Look for pictures or visuals that contain the word.

In the following table, you can find different kinds of context clues that you can use to help you figure out the meanings of new words.

Study Skills

Search for Context Clues

Comparison and contrast	The runner started the race with energy and excitement, but as she crossed the finish line, the *fatigue* and strain showed on her face.	This sentence contrasts the word *fatigue* with energy and compares it to strain. This tells you that someone who is fatigued is strained and has no energy.
Definition and description	Elena is a *biologist*, a scientist who studies living things.	The sentence describes a *biologist* as someone who studies living things.
Synonyms	Carl is very *dependable*. His teachers and his parents know that he is reliable and can be trusted.	The word *dependable* is described by the synonyms reliable and trusted.
Tone and setting	An air of *jubilation* surrounded the members of the science team as they received their medals for first place in the national competition.	The setting of the sentence and the action describe a situation that is positive and full of celebration.
A series of clues	The ocelot, jaguar, and mountain lion are all *endangered* species.	The animals that are mentioned are all rare. This tells you something about the word *endangered*.
Cause and effect	The student group was known for its *boisterous* meetings, so the principal asked extra teachers to monitor the meeting and keep order.	*Boisterous* describes the meetings and tells you that something needs extra supervision.

Study Skills

Break It Down

- Find the root word.
- Write it and ask questions about its meaning.
- Find the affix—the part in front of or after the root word.
- Write it down and use a dictionary to look up its meaning.

public·ize

In this table, you can see how to break words into their roots and affixes.

Word	Root	Affix and Meaning	Meaning
imperfect	perfect	im- (not)	not perfect
semicircle	circle	semi- (half)	half of a circle
teacher	teach	-er (one who)	one who teaches
backward	back	-ward (in the direction of)	to the back
publicize	public	-ize (make)	make public

Remember New Words

- Say the word aloud.
- Write another sentence using the word.
- Make flash cards that include the word and its meaning.
- Review your flash cards to help you remember the meanings of the new words.

Study Skills

C. Learn From Visuals

Tables, graphs, photographs, diagrams, and maps are called visuals. Good readers use all kinds of visuals to help them learn.

Active readers

- find the purpose for the visual they see.
- find information in the visual.
- connect the information they find to what they are studying.

Here's how you can improve your skill in learning from visuals.

When You First Look at a Visual

Scan

- Look at the visual.
- Decide its purpose. Why is it there?
- Find the title.
- Read the caption.

Write

- Write the purpose of the visual. Why is it there?
- Write the key information.
- Write the title of the visual.
- Write the main idea or message.

BIOLOGY: The Dynamics of Life **181**

Study Skills

As You Study the Visual

Graphs

Graphs are pictures of related information. A graph tells you something about a specific situation. There are many kinds of graphs. One of the most common is the bar graph.

Number of Students in Each Grade

A bar graph helps you compare similar information about different items. The separate items being measured are shown as rectangles side by side on the graph.

Diagrams

A diagram is a drawing that has labels on it. It can show how something works or what the parts are called.

Amoeba

A diagram often gives the names of the parts of something, like this diagram of an amoeba. Science books often have many diagrams.

Study Skills

Tables

Tables organize words and numbers for easier reading. They have a title, columns (up and down), and rows (side to side). In this table, the columns show the innings, and the rows show the points each team scored.

Points Earned in the Baseball Game										
Inning	1	2	3	4	5	6	7	8	9	Total Points
Green Team	0	0	1	1	0	0	0	3	0	5
Blue Team	1	0	1	0	2	0	1	0	1	6

Maps

Maps give all kinds of different information. Some examples are location, direction, and land features. They can have words, symbols, numbers, lines, and colors.

Coal Fields of the United States

Figure 6.11

Coal is the most abundant fossil fuel on Earth. The coal deposits of the United States are mainly bituminous coal, which is preferred for electric power generation.

Study Skills

D. Make Chapter and Section Idea Maps

Active readers organize the information they read.

Active readers

- divide the information into smaller units.
- put the information in a logical order.

Starting Out

Scan and Write

- Scan the chapter for main topics and subheadings—in your biology text-book, red headings are main topics and blue headings are subtopics.
- Scan for **boldface** key terms.
- Scan for any visuals.
- Write the information in some kind of graphic map.

Here's an example of one kind of idea map.

Idea Map

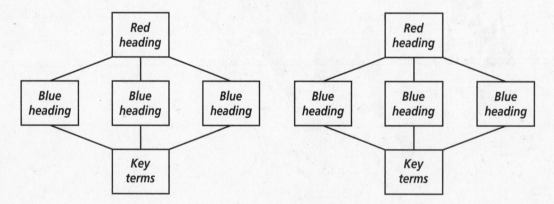